Modern Control Engineering

Modern Control Engineering

Ramona Howell

 Larsen & Keller
www.larsen-keller.com

Modern Control Engineering
Ramona Howell
ISBN: 978-1-64172-425-8 (Hardback)

Larsen & Keller

Published by Larsen and Keller Education,
5 Penn Plaza,
19th Floor,
New York, NY 10001, USA

Cataloging-in-Publication Data

Modern control engineering / Ramona Howell.
    p. cm.
Includes bibliographical references and index.
ISBN 978-1-64172-425-8
1. Automatic control. 2. Automation. 3. Engineering. I. Howell, Ramona.
TJ213 .M63 2020
629.8--dc23

For more information regarding Larsen and Keller Education and its products, please visit the publisher's website www.larsen-keller.com

TABLE OF CONTENTS

It is with great pleasure that I present this book. It has been carefully written after numerous discussions with my peers and other practitioners of the field. I would like to take this opportunity to thank my family and friends who have been extremely supporting at every step in my life.

The engineering discipline, which deals with the application of automatic control theory for designing systems with desired behavior in controlled environments is referred to as control engineering. It uses sensors and detectors for the measurement of output performance of the process, which are being controlled. Such measurements are used to provide corrective feedback that helps to achieve the desired performance. Modern control engineering applies principles of control theory. Control engineering plays an important role in various control systems ranging from simple household washing machines to high-performance fighter aircraft. This book unfolds the innovative aspects of control engineering, which will be crucial for the progress of this field in the future. The topics covered in this extensive book deal with the core aspects of this subject. It is appropriate for students seeking detailed information in this area as well as for experts.

The chapters below are organized to facilitate a comprehensive understanding of the subject:

Chapter – What is Control Engineering?

The engineering discipline that uses automatic control theory to design systems with desired behaviors in control environments is referred to as control engineering. A block diagram is a flowchart used in engineering whereas signal-flow graph is a graph representing system variables and functional connection between nodes. This is an introductory chapter which will introduce briefly all these significant aspects of control engineering.

Chapter – Control Theory

The subfield of mathematics which deals with the control of continuously operating dynamical systems in engineered processes and machines is called control theory. Various methods and branches of control theory include digital control, hybrid control, multivariable control, nonlinear control, robust control, stochastic control, etc. All these methods and branches of control theory have been carefully analyzed in this chapter.

Chapter – Control Systems

Control Systems can be defined as the systems which manage and regulate the behavior of other systems. They can be categorized into open-loop systems, closed-loop systems, feedback systems, negative feedback systems, etc. This chapter has been carefully written to provide an easy understanding of these varied facets of control systems as well as their advantages and disadvantages.

Chapter – Mathematical Concepts

Various mathematical concepts are applied within control engineering. Some of these concepts are complex analysis, differential equations and linear hamiltonian control systems. This chapter closely examines these key mathematical concepts of control engineering to provide an extensive understanding of the subject.

Chapter – Applications

Control engineering has a wide range of applications in many modern automobiles. A few of these applications include aircraft flight control system, control loading system, electronic flight instrument system, fire-control system, intelligent flight control system, guidance system, etc. The topics elaborated in this chapter will help in gaining a better perspective about these applications of control engineering.

Ramona Howell

What is Control Engineering?

The engineering discipline that uses automatic control theory to design systems with desired behaviors in control environments is referred to as control engineering. A block diagram is a flowchart used in engineering whereas signal-flow graph is a graph representing system variables and functional connection between nodes. This is an introductory chapter which will introduce briefly all these significant aspects of control engineering.

Control system engineering is the branch of engineering which deals with the principles of control theory, to design a system which gives yields the desired behavior in a controlled manner. Hence, although control engineering is often taught within electrical engineering at university, it is an interdisciplinary topic.

Control system engineers analyze, design, and optimize complex systems which consist of highly integrated coordination of mechanical, electrical, chemical, metallurgical, electronic or pneumatic elements. Thus control engineering deals with a diverse range of dynamic systems which include human and technological interfacing. These systems are broadly referred to as control systems.

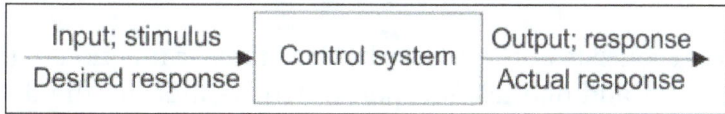

Control system engineering focuses on the analysis and design of systems to improve the speed of response, accuracy, and stability of the system.

The two methods of control system include classical methods and modern methods. The mathematical model of the system is set up as the first step followed by analysis, designing and testing. Necessary conditions for the stability are checked and finally, optimization follows.

In the classical method, mathematical modeling is usually done in the time domain, frequency domain or complex domain. The step response of a system is mathematically modeled in time domain differential analysis to find its settling time, % overshoot, etc. Laplace transforms are most commonly used in the frequency domain to find the open loop gain, phase margin, bandwidth etc of the system. The concept of the transfer function, Nyquist stability criteria, sampling of data, Nyquist plot, poles and zeros, Bode plots, system delays all come under the umbrella of classical control engineering stream.

Modern control engineering deals with Multiple Input Multiple Output (MIMO) systems, State space approach, Eigenvalues, and vectors, etc. Instead of transforming complex ordinary differential equations, modern approach converts higher order equations to first order differential equations and solved by vector method.

Automatic control systems are most commonly used as it does not involve manual control. The controlled variable is measured and compared with a specified value to obtain the desired result. As a result of automated systems for control purposes, the cost of energy or power, as well as the cost of the process, will be reduced increasing its quality and productivity.

BLOCK DIAGRAMS

Block diagrams consist of a single block or a combination of blocks. These are used to represent the control systems in pictorial form.

Basic Elements of Block Diagram

The basic elements of a block diagram are a block, the summing point and the take-off point. Let us consider the block diagram of a closed loop control system as shown in the following figure to identify these elements.

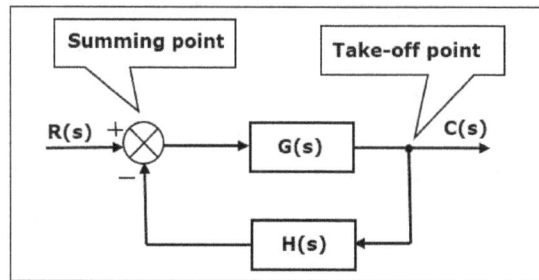

The above block diagram consists of two blocks having transfer functions G(s) and H(s). It is also having one summing point and one take-off point. Arrows indicate the direction of the flow of signals.

Block

The transfer function of a component is represented by a block. Block has single input and single output.

The following figure shows a block having input X(s), output Y(s) and the transfer function G(s).

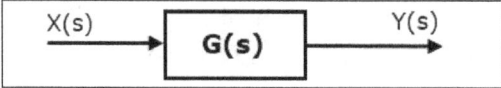

Transfer Function,

$$G(s) = \frac{Y(s)}{X(s)}$$

$$\Rightarrow Y(s) = G(s)X(s)$$

Output of the block is obtained by multiplying transfer function of the block with input.

Summing Point

The summing point is represented with a circle having cross (X) inside it. It has two or more inputs and single output. It produces the algebraic sum of the inputs. It also performs the summation or subtraction or combination of summation and subtraction of the inputs based on the polarity of the inputs. Let us see these three operations one by one.

The following figure shows the summing point with two inputs (A, B) and one output (Y). Here, the inputs A and B have a positive sign. So, the summing point produces the output, Y as sum of A and B.

i.e., $Y = A + B$.

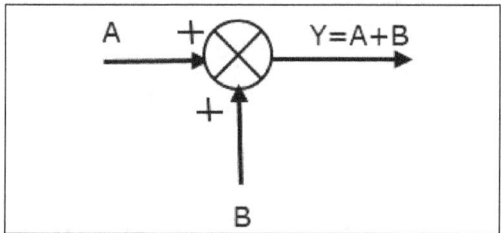

The following figure shows the summing point with two inputs (A, B) and one output (Y). Here, the inputs A and B are having opposite signs, i.e., A is having positive sign and B is having negative sign. So, the summing point produces the output Y as the difference of A and B.

$Y = A + (-B) = A - B$.

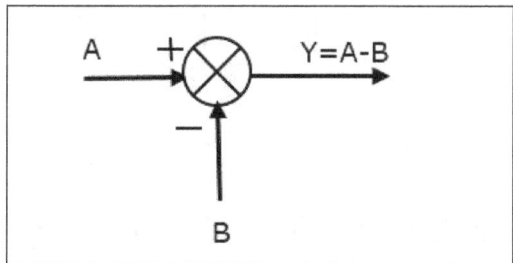

The following figure shows the summing point with three inputs (A, B, C) and one output (Y). Here, the inputs A and B are having positive signs and C is having a negative sign. So, the summing point produces the output Y as

$Y = A + B + (-C) = A + B - C$.

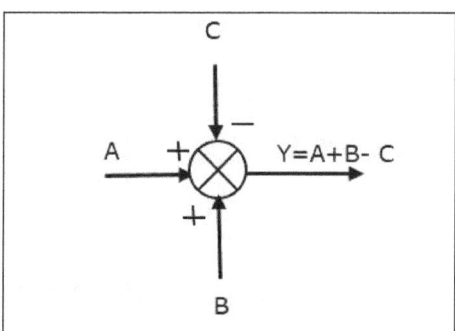

Take-off Point

The take-off point is a point from which the same input signal can be passed through more than one branch. That means with the help of take-off point, we can apply the same input to one or more blocks, summing points.

In the following figure, the take-off point is used to connect the same input, R(s) to two more blocks.

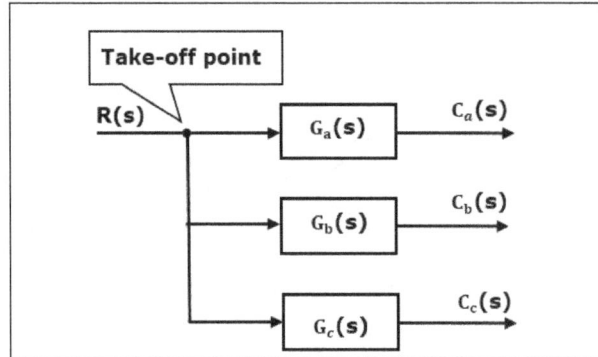

In the following figure, the take-off point is used to connect the output C(s), as one of the inputs to the summing point.

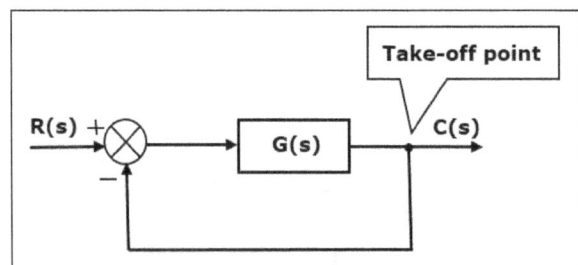

Block Diagram Representation of Electrical Systems

Let us represent an electrical system with a block diagram. Electrical systems contain mainly three basic elements — resistor, inductor and capacitor.

Consider a series of RLC circuit as shown in the following figure. Where, $V_i(t)$ and $V_o(t)$ are the input and output voltages. Let i(t) be the current passing through the circuit. This circuit is in time domain.

By applying the Laplace transform to this circuit, will get the circuit in s-domain. The circuit is as shown in the following figure.

From the above circuit, we can write:

$$I(s) = \frac{V_i(s) - V_o(s)}{R + sL}$$

$$\Rightarrow I(s) = \left\{\frac{1}{R + sL}\right\}\{V_i(s) - V_o(s)\}$$

$$V_o(s) = \left(\frac{1}{sC}\right)I(s)$$

Let us now draw the block diagrams for these two equations individually. And then combine those block diagrams properly in order to get the overall block diagram of series of RLC Circuit (s-domain).

Equation $\Rightarrow I(s) = \left\{\dfrac{1}{R + sL}\right\}\{V_i(s) - V_o(s)\}$ can be implemented with a block having the transfer

function, $\dfrac{1}{R + sL}$. The input and output of this block are $\{V_i(s) - V_o(s)\}$ and $I(s)$. We require a

summing point to get $\{V_i(s) - V_o(s)\}$. The block diagram of Equation 1 is shown in the following figure.

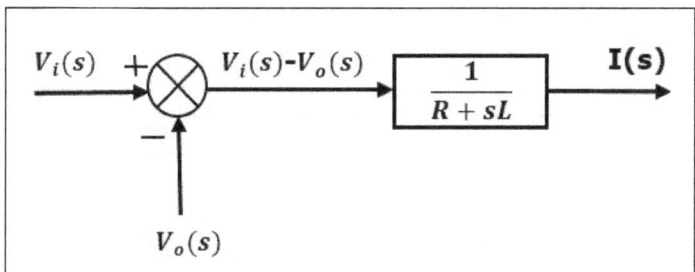

Equation $V_o(s) = \left(\dfrac{1}{sC}\right)I(s)$ can be implemented with a block having transfer function, $\dfrac{1}{sC}$. The

input and output of this block are $I(s)$ and $V_o(s)$. The block diagram of Equation $V_o(s) = \left(\dfrac{1}{sC}\right)I(s)$

is shown in the following figure.

The overall block diagram of the series of RLC Circuit (s-domain) is shown in the following figure.

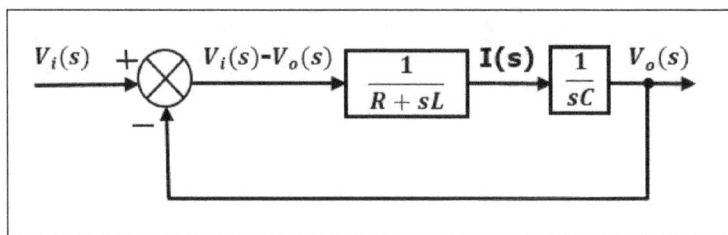

Similarly, you can draw the block diagram of any electrical circuit or system just by following this simple procedure.

- Convert the time domain electrical circuit into an s-domain electrical circuit by applying Laplace transform.

- Write down the equations for the current passing through all series branch elements and voltage across all shunt branches.

- Draw the block diagrams for all the above equations individually.

- Combine all these block diagrams properly in order to get the overall block diagram of the electrical circuit (s-domain).

Steps to Draw Block Diagram

Consider a simple R-L circuit,

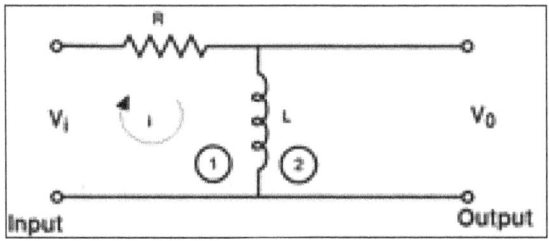

Apply KVL,

$$V_i = R_i + L\frac{di}{dt}$$

$$V_0 = L\frac{di}{dt}$$

Now taking laplace transform of Eq. $V_i = R_i + L\dfrac{di}{dt}$ and Eq. $V_0 = L\dfrac{di}{dt}$ with initial condition zero

$$V_i(s) = I(s)R + SLI(s)$$

$$V_i(s) = I(s)(R + SL)$$

$$V_0(s) = SLI(s)$$

From eq. $V_i(s) = I(s)(R + SL)$ and eq. $V_0(s) = SLI(s)$

$$\frac{V_0(s)}{V_i(s)} = \frac{sL}{R + SL}$$

From figure:

$$i = \frac{V_i - V_0}{R}$$

$$V_0 = L\frac{di}{dt}$$

Now taking laplace transform of Eq. $i = \dfrac{V_i - V_0}{R}$, and Eq. $V_0 = L\dfrac{di}{dt}$

$$I(s) = \frac{1}{R}\big[V_i(s) - V_0(s)\big]$$

$$V_0(s) = SLI(s)$$

For the right-hand side of eq. $i = \dfrac{V_i - V_0}{R}$, we will use a summing point.

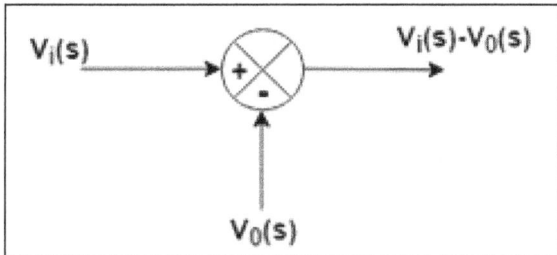

Here the output of summing point is given to the block, and the output of the block is I(s)

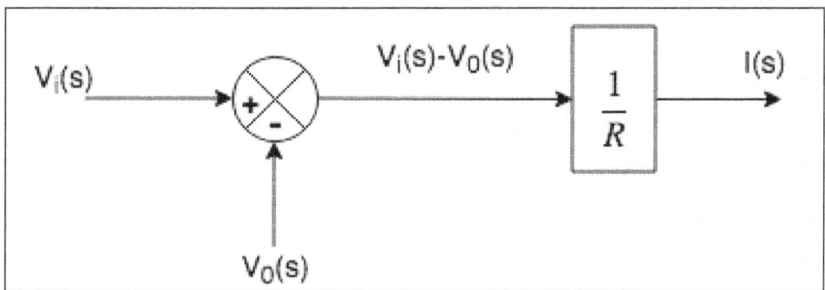

Now the output I(s) is given to another block containing element SL and the output of this block is V_0.

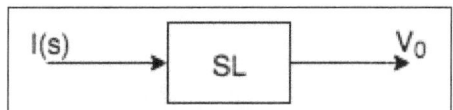

By combining the above two figures, we get the required block diagram.

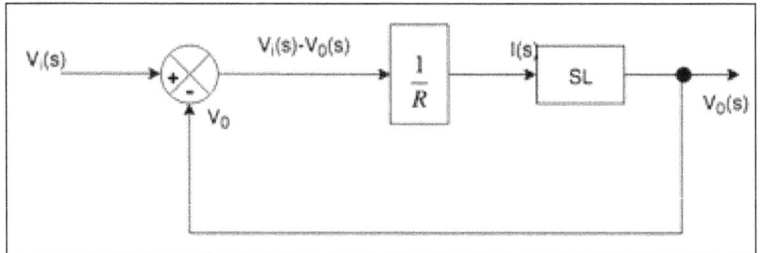

Closed Loop Control System

A system in which a feedback path is there is called a closed-loop control system. In this system, the output is feedback into the error detector and then it is compared with the input signal. The feedback signal can be negative or positive.

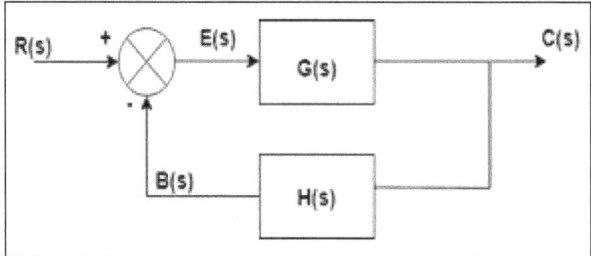

For positive feedback

$$\frac{C(s)}{R(s)} = \frac{G(s)}{1+G(s)H(s)}$$

And for negative feedback

$$\frac{C(s)}{R(s)} = \frac{G(s)}{1+G(s)H(s)}$$

Block diagram reduction rules:

Rule No.1: Blocks in Cascade

When two or more blocks are connected in series, then the resultant block is the product of the individual blocks.

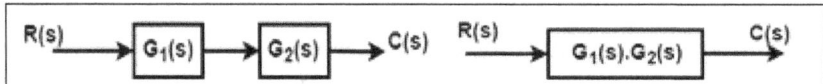

Rule No.2: Blocks in Parallel

When two or more blocks are connected in parallel, then the resultant block is the sum of the individual blocks.

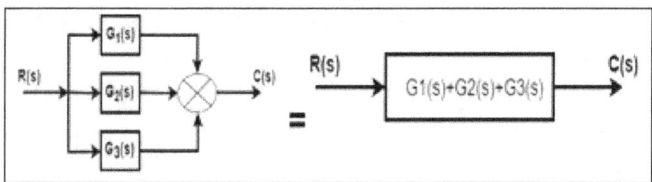

Rule No.3: Moving a Take-off Point Ahead of a Block

When the take-off point is moved ahead of a block (before the block), then the same transfer function is introduced in the take-off point branch.

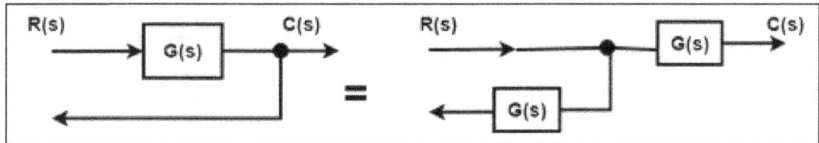

Rule No.4: Moving the Take-off Point after the Block

When the take-off point is moved after the block, then a block with reciprocal of a transfer function is introduced in the take-off point branch.

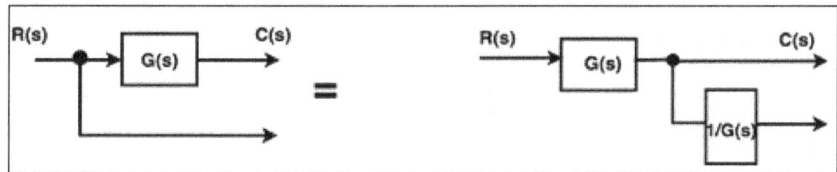

Rule No.5: Moving a Summing Point Beyond the Block

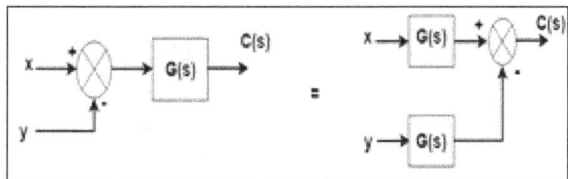

Rule No.6: Moving a Summing Point Ahead of a Block

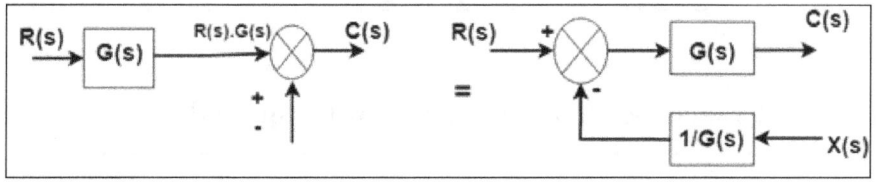

Rule No.7: Interchanging two Summing Points

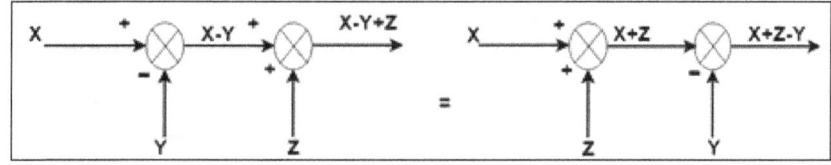

Rule No.8: Moving a Take-off Point beyond a Summing Point

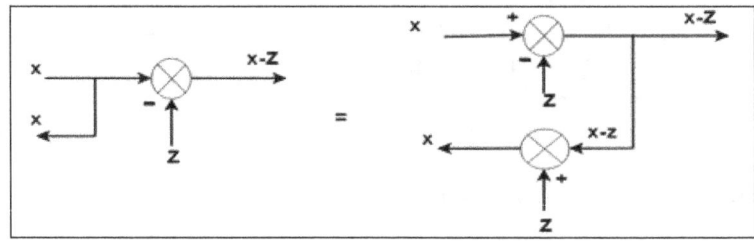

Rule No.9: Moving a Take-off Point ahead of a Summing Point

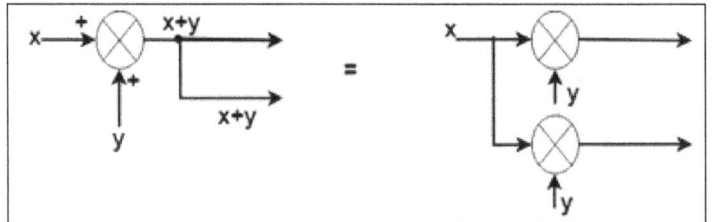

Rule No.10: Eliminating a Forward Loop

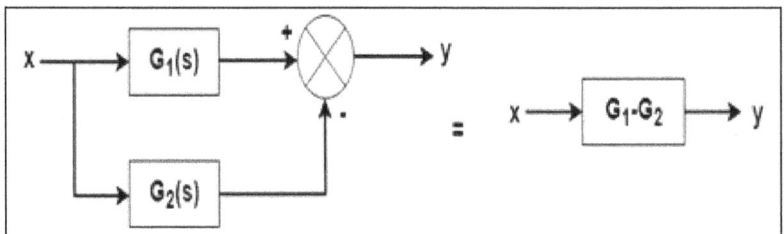

Example:

Find the transfer function of the following by block reduction technique.

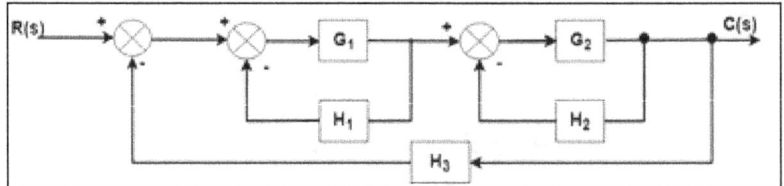

Solution:

Step 1: There are two internal closed loops. Firstly, we will remove this loop.

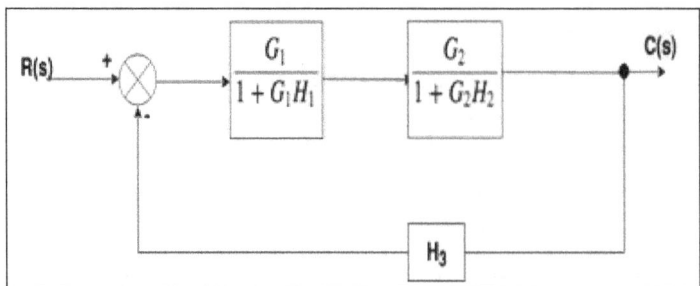

Step 2: When the two blocks are in a cascade or series we will use rule no.1.

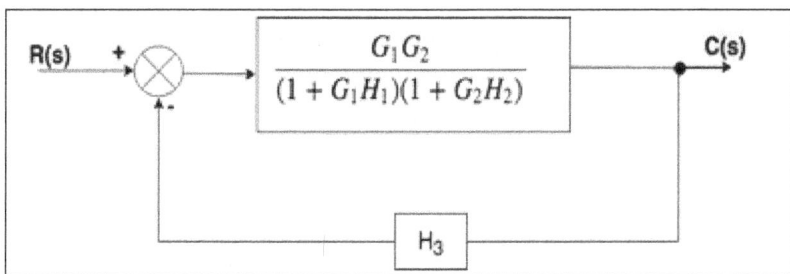

Step 3: Now we will solve this loop.

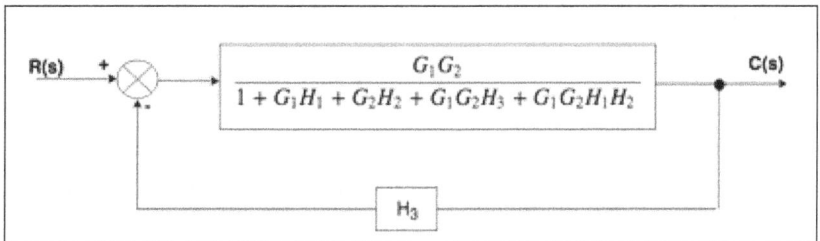

Step 4: Ans .

$$\frac{C(s)}{R(s)} = \frac{G_1 G_2}{1 + G_1 H_1 + G_2 H_2 + G_1 G_2 H_3 + G_1 G_2 H_1 H_2}$$

SIGNAL FLOW GRAPHS

Signal flow graph is a graphical representation of algebraic equations.

Basic Elements of Signal Flow Graph

Nodes and branches are the basic elements of signal flow graph.

Node

Node is a point which represents either a variable or a signal. There are three types of nodes — input node, output node and mixed node.

- Input Node: It is a node, which has only outgoing branches.

- Output Node: It is a node, which has only incoming branches.

- Mixed Node: It is a node, which has both incoming and outgoing branches.

Example:

Let us consider the following signal flow graph to identify these nodes.

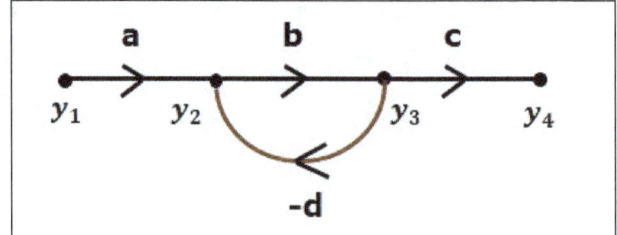

- The nodes present in this signal flow graph are y_1, y_2, y_3 and y_4.

- y1 and y4 are the input node and output node respectively.

- y_2 and y_3 are mixed nodes.

Branch

Branch is a line segment which joins two nodes. It has both gain and direction. For example, there are four branches in the above signal flow graph. These branches have gains of a, b, c and -d.

Construction of Signal Flow Graph

Let us construct a signal flow graph by considering the following algebraic equations –

$$y_2 = a_{12}y_1 + a_{42}y_4$$

$$y_3 = a_{23}y_2 + a_{53}y_5$$

$$y_4 = a_{34}y_3$$

$$y_5 = a_{45}y_4 + a_{35}y_3$$

$$y_6 = a_{56}y_5$$

There will be six nodes (y_1, y_2, y_3, y_4, y_5 and y_6) and eight branches in this signal flow graph. The gains of the branches are a_{12}, a_{23}, a_{34}, a_{45}, a_{56}, a_{42}, a_{53} and a_{35}.

To get the overall signal flow graph, draw the signal flow graph for each equation, then combine all these signal flow graphs and then follow the steps given below –

Step 1: Signal flow graph for $y_2 = a_{13}y_1 + a_{42}y_4$ is shown in the following figure.

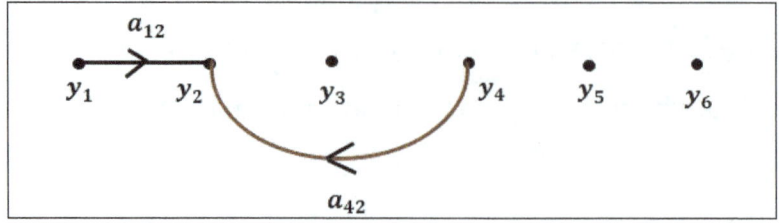

Step 2: Signal flow graph for $y_3 = a_{23}y_2 + a_{53}y_5$ is shown in the following figure.

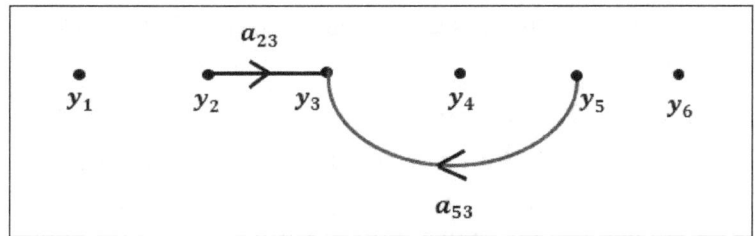

Step 3: Signal flow graph for $y_4 = a_{34}y_3$ is shown in the following figure.

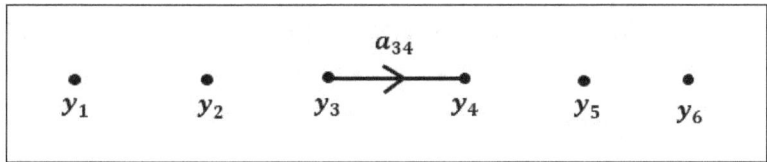

Step 4: Signal flow graph for $y_5 = a_{45}y_4 + a_{35}y_3$ is shown in the following figure.

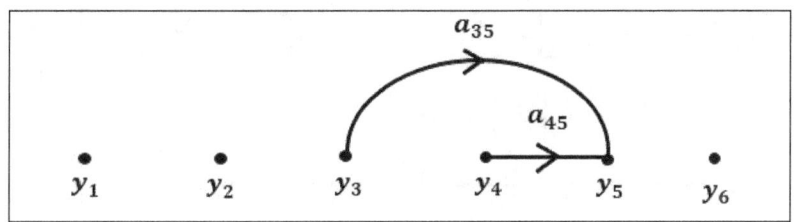

Step 5: Signal flow graph for $y_6 = a_{56}y_5$ is shown in the following figure.

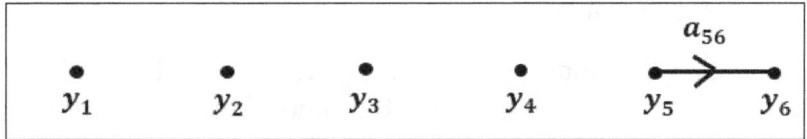

Step 6: Signal flow graph of overall system is shown in the following figure.

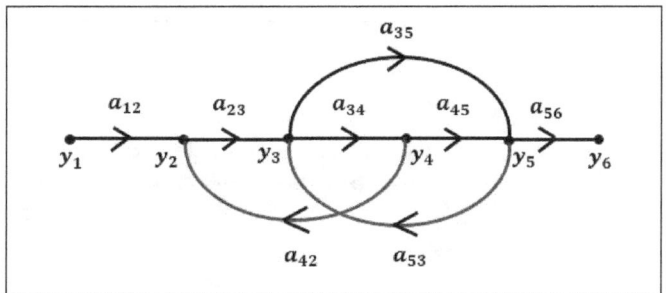

Conversion of Block Diagrams into Signal Flow Graphs

Follow these steps for converting a block diagram into its equivalent signal flow graph.

- Represent all the signals, variables, summing points and take-off points of block diagram as nodes in signal flow graph.

- Represent the blocks of block diagram as branches in signal flow graph.

- Represent the transfer functions inside the blocks of block diagram as gains of the branches in signal flow graph.

- Connect the nodes as per the block diagram. If there is connection between two nodes (but there is no block in between), then represent the gain of the branch as one. For example, between summing points, between summing point and takeoff point, between input and summing point, between take-off point and output.

Example:

Let us convert the following block diagram into its equivalent signal flow graph.

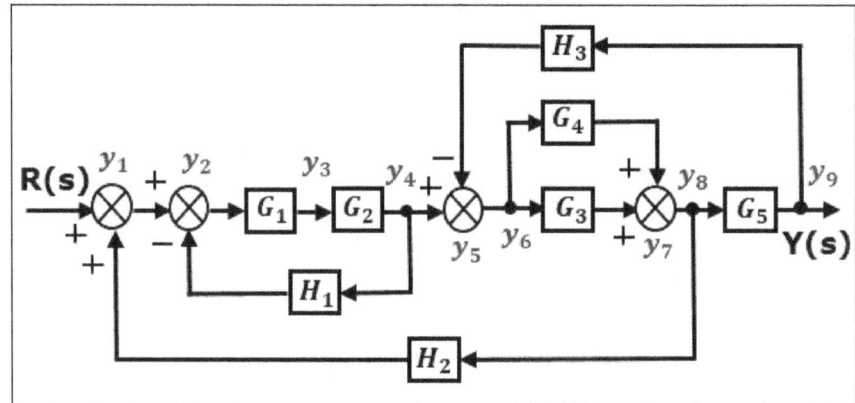

Represent the input signal R(s) and output signal C(s) of block diagram as input node R(s)and output node C(s) of signal flow graph.

Just for reference, the remaining nodes (y_1 to y_9) are labelled in the block diagram. There are nine nodes other than input and output nodes. That is four nodes for four summing points, four nodes for four take-off points and one node for the variable between blocks G_1 and G_2.

The following figure shows the equivalent signal flow graph.

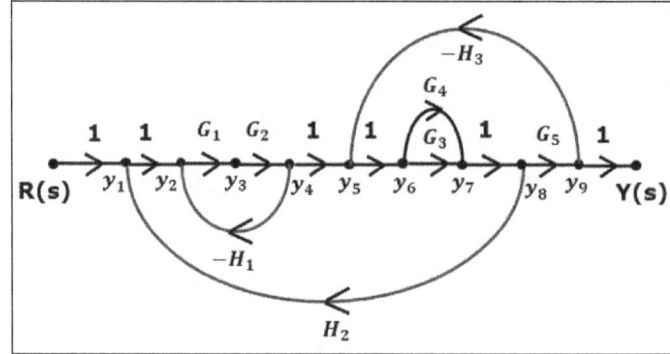

With the help of Mason's gain formula, you can calculate the transfer function of this signal flow graph. This is the advantage of signal flow graphs. Here, we no need to simplify (reduce) the signal flow graphs for calculating the transfer function.

DIGITAL DATA OF CONTROL SYSTEM

So let us discuss first some advantages of digital system over analog system.

- Power consumption is less in digital system as compared to analog system.

- Digital systems can handle non linear system easily which is the most important advantage of digital data in control system.

- Digital systems works on the logical operations due to this they show decision making property which is very useful in the present world of machines.

- They are more reliable as compared to analog systems.

- Digital systems are easily available in compact size and have light weight.

- They works on instructions we can program them as per our needs hence we can they are more versatile than analog systems.

- Various complex tasks can be performed easily by the help of digital technology with a high degree of accuracy.

Sampling Process

Sampling process is defined as the conversion of analog signal into the digital signal with the help of a switch (also known as sampler). A sampler is a continuous ON and OFF switch which directly converts analog signals into digital signals. We may have a series connection of sampler depending upon the conversion of signals we use them. For an ideal sampler, the width of the output pulse is very small (tending to zero). Now when talk about discrete system it is very important to know about the z transformations. We will discuss here about the z transformations and its utilities in discrete system. Role of z transformation in discrete systems is same as Fourier transform in continuous systems. Now let us discuss z transformation in detail.

We define z transform as,

$$f(z) = \sum_{k=-\infty}^{\infty} f(k)z^{-k}$$

Where, $F(k)$ is a discrete data,

Z is a complex number,

$F(z)$ is Fourier transform of $f(k)$.

Important Properties of z transformation are written below:

Linearity

Let us consider summation of two discrete functions $f(k)$ and $g(k)$ such that

$$p \times f(k) + qg(k)$$

such that p and q are constants, now on taking the Laplace transform we have by property of linearity:

$$Z[p \times f(k) + q \times g(k)] = p \times Z[f(k)] + q \times Z[g(k)]$$

Change of Scale: let us consider a function f(k), on taking the z transform we have,

$$Z[f(k)] = f(z)$$

then we have by change of scale property,

$$z[a^k f(k)] = f\left(\frac{z}{a}\right)$$

Shifting Property

As per this property,

$$Z[F(k + or - n)] = z^{(+ or - n)} f(z)$$

Now let us discuss some important z transforms.

  Function $f(t) = t$

Laplace transformation of this function is $1/s^2$ and the corresponding f(k) = kT. Now the z transformation of this function is:

$$\frac{Tz}{(z-1)^2}$$

Function f (t) = t²: Laplace transformation of this function is $2/s^3$ and the corresponding f(k) = kT. Now the z transformation of this function is:

$$T^2 \times z \times \frac{z+1}{(z-1)^3}$$

  Function $f(t) = e^{-at}$

Laplace transformation of this function is $1/(s + a)$ and the corresponding f(k) = $e^{(-akT)}$. Now the z transformation of this function is:

$$\frac{z}{z - e^{aT}}$$

  Function $f(t) = te^{-at}$

Laplace transformation of this function is $1/(s + a)^2$ and the corresponding f(k) = Te^{-akT}. Now the z transformation of this function is:

$$T\frac{e^{-aT}z}{\left(z-e^{-aT}\right)^2}$$

Function $f(t) = sin(at)$

Laplace transformation of this function is $a/(s^2 + a^2)$ and the corresponding $f(k) = sin(akT)$. Now the z transformation of this function is:

$$\frac{sin(aT)z}{z^2 - 2cos(aT)z + 1}$$

Function $f(t) = cos(at)$

Laplace transformation of this function is $s/(s^2 + a^2)$ and the corresponding $f(k) = cos(akT)$. Now the z transformation of this function is:

$$\frac{z^2 - cos(aT)z}{z^2 - 2cos(aT)z + 1}$$

Now sometime there is a need to sample data again, which means converting discrete data into continuous form. We can convert digital data of control system into continuous form by hold circuits.

Hold Circuits

These are the circuits which converts discrete data into continuous data or original data. Now there are two types of hold circuits:

Zero Order Hold Circuit: The block diagram representation of the zero order hold circuit is given below-

In the block diagram we have given an input f(t) to the circuit, when we allow input signal to pass through this circuit it reconverts the input signal into continuous one. The output of the zero order hold circuit is shown below.

Now we are interested in finding out the transfer function of the zero order hold circuit. On writing the output equation we have:

$$F_o(t) = u(t) - u(t - T)$$

on taking the Laplace transform of the above equation we have,

$$F_0(s) = \frac{1}{s - \dfrac{1}{s \cdot e^{-sT}}}$$

From the above equation we can calculate transfer function as,

$$\frac{1 - e^{-sT}}{s}$$

On substituting s=jω we can draw the bode plot for the zero order hold circuit. The electrical representation of the zero order hold circuit is shown below, which consists of a sampler connected in series with a resistor and this combination is connected with a parallel combination of resistor and capacitor.

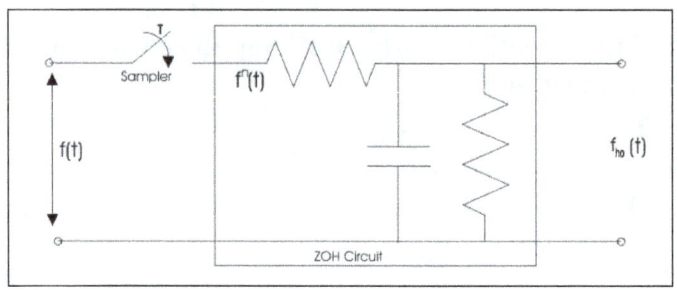

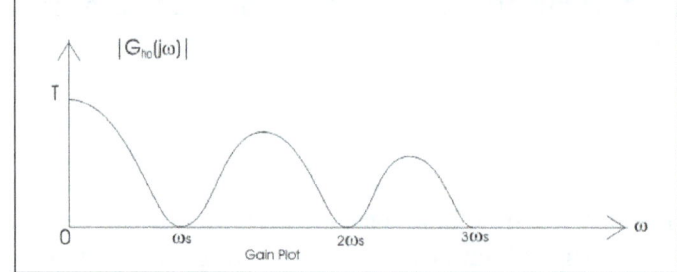

GAIN PLOT – frequency response curve of ZOH.

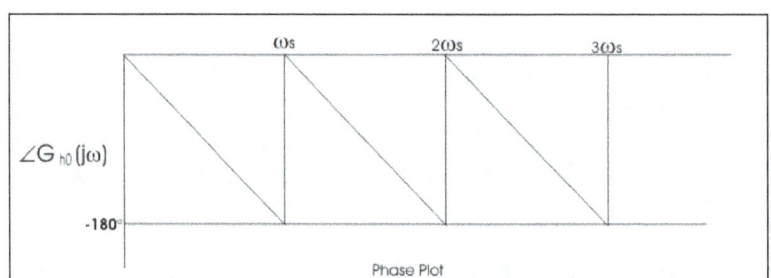

PHASE PLOT – frequency response curve of ZOH.

First Order Hold Circuit: The block diagram representation of the first order hold circuit is given below:

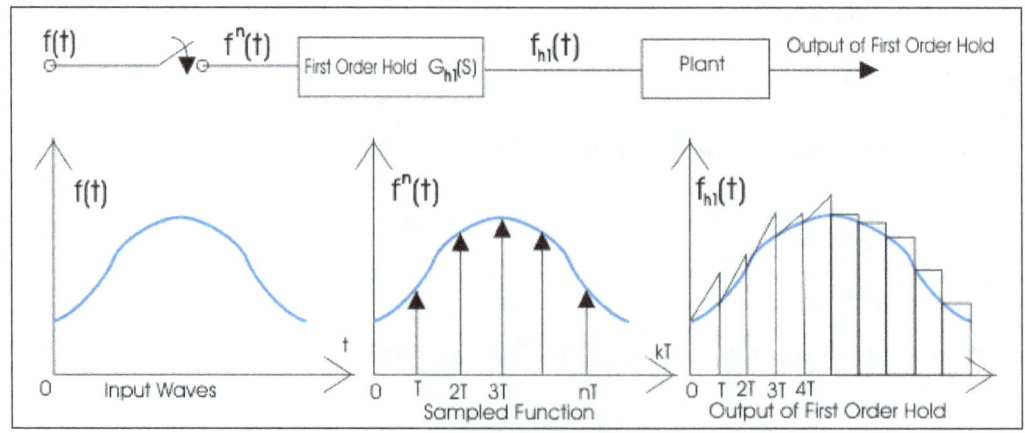

First Order Hold Circuit.

In the block diagram we have given an input f(t) to the circuit, when we allow input signal to pass through this circuit it reconverts the input signal into continuous one. The output of the first order hold circuit is shown below: Now we are interested in finding out the transfer function of the first order hold circuit. On writing the output equation we have:

$$F_o(t) = \left(1 + \frac{t}{T}\right)u(t) - \frac{t-T}{T}u(t-1) - u(t-1)$$

On taking the Laplace transform of the above equation we have:

$$F_o(s) = \frac{(1 - e^{-Ts}) \times (Ts + 1)}{Ts^2}$$

From the above equation we can calculate transfer function as $\left(1 - e^{-sT}\right)/$. on substituting $s = j\omega$ we can draw the bode plot for the zero order hold circuit. The bode plot for the first order hold circuit is shown below which consists of a magnitude plot and a phase angle plot. The magnitude plot starts with magnitude value $2\pi / \omega_s$.

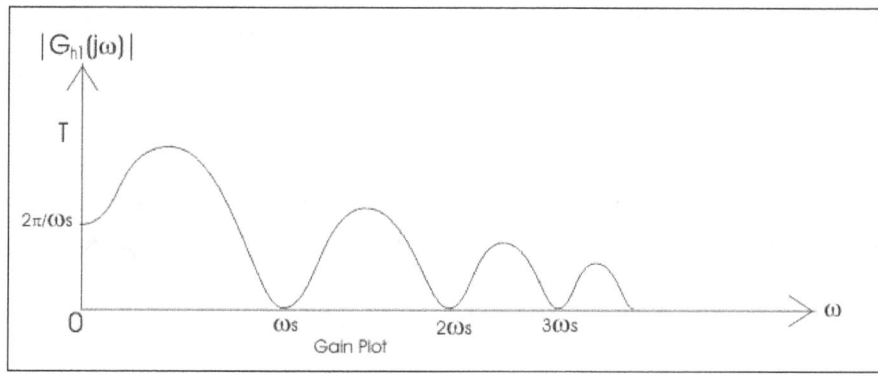

Gain Plot of First Order Hold Circuit.

TIME DOMAIN ANALYSIS OF CONTROL SYSTEM

In a control system, there may be some energy storing elements attached to it. Energy storing elements are generally inductors and capacitors in case of an electrical system. Due to the presence of these energy storing elements, if the energy state of the system is disturbed, it will take a certain time to change from one energy state to another. The exact time taken by the system for changing one energy state to another is known as transient time and the value and pattern voltages and currents during this period are known as the transient response.

A transient response is normally associated with an oscillation, which may be sustained or decaying in nature. The exact nature of the system depends upon the parameters of the system. Any system can be represented with a linear differential equation. The solution of this linear differential equation gives the response of the system. The representation of a control system by a linear differential equation of functions of time and its solution is collectively called time domain analysis of the control system.

Step Function

Let us take an independent voltage source or a battery which is connected across a voltmeter via a switch, s. It is clear from the figure below, whenever the switch s is open, the voltage appears between the voltmeter terminals is zero. If the voltage between the voltmeter terminals is represented as $v(t)$, the situation can be mathematically represented as,

$$v(t) = 0 \ when -\infty < t < 0$$

Now let us consider at t = 0, the switch is closed and instantly the battery voltage V volt appears across the voltmeter and that situation can be represented as,

$$v(t) = k \ when \ 0 < t < \infty$$

Combining the above two equations we get

$$v(t) = 0 \quad when \ -\infty < t < 0$$
$$and \qquad = k \quad when \quad 0 < t < \infty$$

In the above equations if we put 1 in place of V, we will get a unit step function which can be defined as

$$u(t) = 0 \quad when \quad t \le 0$$
$$and \qquad = 1 \quad when \quad t \ge 0$$

Now let us examine the Laplace transform of unit step function. Laplace transform of any function can be obtained by multiplying this function by e^{-st} and integrating multiplied from 0 to infinity.

$$\mathcal{L}u(t) = \int_0^\infty u(t)e^{-st}dt = \int_0^\infty 1 . e^{-st}dt = \left[\frac{e^{-st}}{-s}\right]_0^\infty = \frac{1}{s}$$

If input is R(s), then,

$$R(s) = \frac{1}{s}$$

Ramp Function

The function which is represented by an inclined straight line intersecting the origin is known as ramp function. That means this function starts from zero and increases or decreases linearly with time. A ramp function can be represented as,

$$r(t) = 0 \quad when \ t < 0$$
$$and \qquad = kt \quad when \quad t > 0$$

Here in this above equation, k is the slope of the line.

Now let us examine the Laplace transform of ramp function. Laplace transform of any function can be obtained by multiplying this function by e^{-st} and integrating multiplied from 0 to infinity.

$$\mathcal{L}r(t) = \int_0^\infty r(t)e^{-st}dt = \int_0^\infty kt \cdot e^{-st} \, dt = \frac{k}{s^2}$$

$$R(S) = \frac{k}{s^2}$$

Parabolic Function

Here, the value of function is zero when time t<0 and is quadratic when time t > 0. A parabolic function can be defined as,

$$p(t) = 0 \quad when \ t < 0$$

$$and \qquad = \frac{kt^2}{2} \quad when \ t > 0$$

Now let us examine the Laplace transform of parabolic function. Laplace transform of any function can be obtained by multiplying this function by e^{-st} and integrating multiplied from 0 to infinity.

$$\mathcal{L}p(t) = \int_0^\infty p(t)e^{-st}dt = \int_0^\infty \frac{kt^2}{2} \cdot e^{-st}dt = \frac{k}{s^3}$$

$$R(S) = \frac{k}{s^3}$$

Impulse Function

Impulse signal is produced when input is suddenly applied to the system for infinitesimal duration of time. The waveform of such signal is represented as impulse function. If the magnitude of such function is unity, then the function is called unit impulse function. The first time derivative of step function is impulse function. Hence Laplace transform of unit impulse function is nothing but Laplace transform of first-time derivative of unit step function.

$$\mathcal{L} \ (Unit \ impulse \ function) = \mathcal{L}\frac{d}{dt}(Unit \ step \ function)$$

$$= s\mathcal{L} \ (Unit \ step \ function) = s.\frac{1}{s} = 1$$

Time Response of First Order Control Systems

When the maximum power of s in the denominator of a transfer function is one, the transfer function represents a first order control system. Commonly, the first order control system can be represented as:

$$\frac{C(s)}{R(s)} = \frac{1}{sT+1}$$

Time Response for Step Function

Now a unit step input is given to the system, then let us analyze the expression of the output:

$$\frac{C(s)}{R(s)} = \frac{1}{sT+1} \Rightarrow C(s) = R(s)\frac{1}{sT+1}$$

$$\Rightarrow C(s) = \frac{1}{s} \cdot \frac{1}{sT+1}\left[\because R(s) = \frac{1}{s}\right] = \frac{T}{sT(sT+1)}$$

$$= T\left[\frac{sT+1-sT}{sT(sT+1)}\right] = T\left[\frac{1}{sT} - \frac{1}{(sT+1)}\right]$$

$$= \frac{1}{s} - \frac{T}{sT+1} = \frac{1}{s} - \frac{1}{s+\dfrac{1}{T}}$$

$$\therefore \mathcal{L}^{-1}[C(s)] = \mathcal{L}^{-1}\left[\frac{1}{s} - \frac{1}{s+\dfrac{1}{T}}\right]$$

$$\Rightarrow c(t) = \mathcal{L}^{-1}\left[\frac{1}{s}\right] - \mathcal{L}^{-1}\left[\frac{1}{s+\dfrac{1}{T}}\right] = 1 - e^{-t/T}$$

$$\left(\because \mathcal{L}^{-1}\left[\frac{1}{s}\right] = 1 \text{ and } \mathcal{L}^{-1}\left[\frac{1}{s+a}\right] = e^{-at}\right)$$

It is seen from the error equation that if the time approaching to infinity, the output signal reaches exponentially to the steady-state value of one unit. As the output is approaching towards input exponentially, the steady-state error is zero when time approaches to infinity.

$$Error\ e(t) = r(t) - c(t)$$

$$\therefore Steady\ State\ Error$$

$$= \lim_{t \to \infty}\left[1 - \left(1 - e^{-t/T}\right)\right] = \lim_{t \to \infty} e^{-t/T} = 0$$

Let us put t = T in the output equation and then we get,

$$c(T) = 1 - e^{-T/T} = 1 - e^{-1} = 1 - 0.368 = 0.632$$

This T is defined as the time constant of the response and the time constant of a response signal is that time for which the signal reaches to its 63.2 % of its final value. Now if we put t = 4T in the above output response equation, then we get,

$$c(T) = 1 - e^{-4T/T} = 1 - e^{-4} = 1 - 0.018 = 0.982$$

When the actual value of the response reaches to the 98% of the desired value, then the signal is said to be reached to its steady-state condition. This required time for reaching the signal to 98 % of its desired value is known as setting time and naturally setting time is four times of the time constant of the response. The condition of response before setting time is known as transient condition and condition of the response after setting time is known as steady-state condition. From this explanation, it is clear that if the time constant of the system is smaller, the response of the system reaches its steady-state condition faster.

Time Response for Ramp Function

$$\frac{C(s)}{R(s)} = \frac{1}{sT+1} \Rightarrow C(s) = R(s)\frac{1}{sT+1}$$

$$\Rightarrow C(s) = \frac{1}{s^2} \cdot \frac{1}{sT+1} \left[\because R(s) = \frac{1}{s^2} \right]$$

$$= \frac{1 - s^2T^2 + s^2T^2}{s^2(sT+1)} = \frac{(1+sT)(1-sT)}{s^2(sT+1)} + \frac{s^2T^2}{s^2(sT+1)}$$

$$= \frac{(1-sT)}{s^2} + \frac{T^2}{(sT+1)} = \frac{1}{s^2} - \frac{T}{s} + \frac{T^2}{(sT+1)}$$

$$= \frac{1}{s^2} - \frac{T}{s} + \frac{T}{\left(s + \dfrac{1}{T}\right)}$$

$$\therefore c(t) = \mathcal{L}^{-1}[C(s)] = \mathcal{L}^{-1}\left[\frac{1}{s^2} - \frac{T}{s} + \frac{T}{\left(s + \dfrac{1}{T}\right)} \right]$$

$$= \mathcal{L}^{-1}\left[\frac{1}{s^2}\right] - \mathcal{L}^{-1}\left[\frac{T}{s}\right] + \mathcal{L}^{-1}\left[\frac{T}{\left(s + \dfrac{1}{T}\right)} \right]$$

$$= t - T + Te^{-t/T}$$

Error $e(t) = r(t) - c(t)$

$\therefore$ Steady State Error

$$= \lim_{t \to \infty}\left[t - \left(t - T + T\,e^{-t/T}\right) \right] = \lim_{t \to \infty}\left(T - T\,e^{-t/T}\right)$$

$$= \lim_{t \to \infty} T\left(1 - e^{-t/T}\right) = T$$

In this case, during the steady-state condition, the output signal lags behind the input signal by a time equal to the time constant of the system. If the time constant of the system is smaller, the positional error of the response becomes lesser.

Time Response for Impulse Function

$$\frac{C(s)}{R(s)} = \frac{1}{sT+1} \Rightarrow C(s) = R(s)\frac{1}{sT+1}$$

$$\Rightarrow C(s) = 1 \cdot \frac{1}{sT+1} \left[\because R(s) = 1 \right]$$

$$= \frac{1}{T} \cdot \frac{1}{s+\dfrac{1}{T}}$$

$$\therefore c(t) = \mathcal{L}^{-1}[C(s)] = \mathcal{L}^{-1}\left[\frac{1}{T} \cdot \frac{1}{s+\dfrac{1}{T}} \right] = \frac{1}{T}e^{-1/T}$$

In the above explanation of time response of the control system, we have seen that the step function is the first derivative of ramp function and the impulse function is the first derivative of a step function. It is also found that the time response of step function is the first derivative of time response of ramp function and time response of impulse function is the first derivative of time response of step function.

$$\frac{d}{dt}(t-T+Te^{-t/T}) = 1-0+\frac{-T}{T}e^{-t/T} = 1-e^{-t/T}$$

$$and\ \frac{d}{dt}(1-e^{-t/T}) = -\frac{1}{T}(-e^{-t/T}) = \frac{1}{T}e^{-t/T}$$

CONTROL SYSTEMS IN OUR DAILY LIFE

Control System is used to control position, velocity, and acceleration is very common in industrial and military applications. They have been given the special name of servomechanisms. With all their many advantages, CS in advertently act as an oscillator. Through proper design, however, all the advantages of CS can be utilized without having an unstable system.

Several characteristics of CS can be linked to human behavior. CS can "think" in the sense that they can replace to some extent, human operation. CS can distinguish between open-loop and closed-loop CS and it is a concept or principle that seems to fundamental in nature and not necessarily peculiar to engineering. In human social and political organizations, for example, a leader remains the leader only as long as she is successful in realizing the desires of the group. CS theory can be discussed from four viewpoints as: an intellectual discipline within science and the philosophy of science, a part of engineering, with industrial applications and social problems of the present and the future. In global communication, developed countries and developing countries should build several attractive and sound symbiosis bridges, to prevent loss of universe balances. CS applications have social impacts not only in developed countries but also in developing countries.

Human Control Systems (CS)

The relation between the behavior of living creatures and the functioning of CS has recently gained wide attention. Wiener implied that all systems, living and mechanical are both information and CS. Wiener suggested that the most promising techniques for studying both systems are Information theory and CS theory.

Several characteristics of CS can be linked to human behavior. CS can "think" in the sense that they can replace to some extent, human operation. These devices do not have the privilege of freedom in their thinking process and are constrained by the designer to some predetermined function. Adaptive CS, which is capable of modifying their functioning in order to archive optimum performance in a varying environment, have recently gained wide attention. These systems are a step closer to the adaptive capability of human behavior.

The human body is, indeed, a very complex and highly perfected adaptive CS. Consider, for example, the human actions required to steer an automobile. The driver's object is to keep the automobile traveling in the center of a chosen lane on the road. Changes in the direction of the road are compensated for by the driver turning the steering wheel. The driver's object is to keep the input (the car's desired position on the road) and the input (the car's desired position on the road) as close to zero as possible.

Figure below illustrates the block diagram of the CS involved in steering an automobile. The error detector in this case is the brain of the driver. This in turn activates the driver's muscles, which control the steering wheel. Power amplification is provided by the automobile's steering mechanism, which controls the position of the wheels. The feedback element represents the human's sensors (visual and tactile). Of course, this description in very crude, any attempt to construct a mathematical model of the process should somehow account for the adaptability of the human being and the effects of learning, fatigue, motivation, and familiarity with the road.

CS process as that found in physical, biological, and social systems. Many systems control themselves through information feedback, which shows deviations from standards and initiates changes. In other words, systems use some of their energy to feedback information that compares performance with a standards and initiates corrective action.

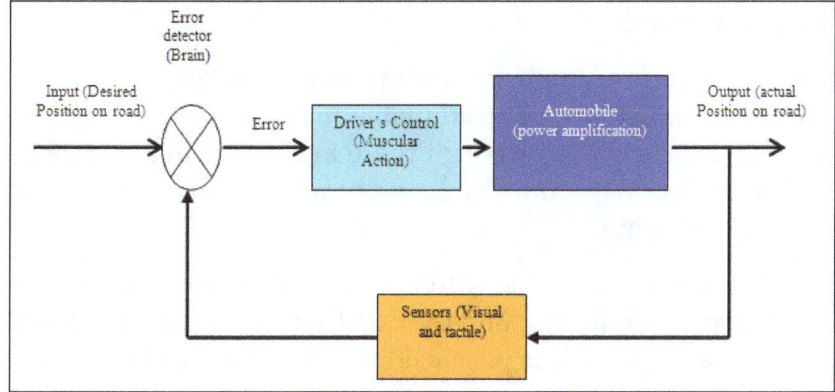

Steering of an automobile: A feedback control system involving human capability.

The house thermostat is a system of feedback and information control. When the house temperature falls below the preset level, an electric message is sent to the heating system, which is then activated. When the temperature increases and reaches the set level, another message shut off the heater. This continual measurement and turning on and off the heater keeps the house at the desired temperature. A similar process activates the air-conditioning system. As soon as the temperature exceeds the preset level, the air-conditioning system cools the house to the desired temperature. Likewise, in the human body, a number of CS control temperature, blood pressure, motor reactions, and other conditions. Another example of feedback is the grade a student receives on a midterm test. This is intended, of course, to give the student information about how he or she is doing and, if performance is less than desirable, to send a signal suggesting improvement.

Control Systems (CS) in Physical Systems

CS is to be found in almost every aspect of our daily environment. In the home, the refrigerator utilizes a temperature-control system. The desired temperature is set and a thermostat measures the actual temperature and the error. A compressor motor is utilized for power amplification. Other applications of control in the home are the hot-water heater, the central heating system, and the oven, which all work on a similar principle. We also encounter CS when driving our automobile. CS is used for maintaining constant speed (cruise control), constant temperature (climate control), steering, suspension, engine control, and to control skidding (antiskid system).

In industry, the term automation is very common. Modern industrial plants utilized robots for manufacturing temperature controls, pressure controls, speed controls, position controls, etc. The chemical process control field is an area where automations have played an important role. Here, the CS engineer is interested in controlling temperature, pressure, humidity, thickness, volume, quality, and many other variables. Areas of additional interest include automatic warehousing, inventory control and automation of farming.

Here, it is presented the state of the CS field by illustrating its application in the following important aspects of engineering: robotics, space travel, commercial rail and air transportation, military systems, surface effect ships, hydrofoils and biomedical CS.

CS in Industrial Robots

A new work force strategy without denying the existingof CS is established by retooling the work forces, thus the challenges of socialimpacts could beanswers wisely and would be bright opportunities toimprove human standards of living.

In manufacturing plants in several countries, there has been a large-scale increase in the usage of CS for industrial robots, which are programmable machine tools designed in many cases to accomplish arduous or complex tasks. Although there has been some opposition to the fact that robots often replace human labor, but the trend toward robotics will continue, and on balance, be beneficial to the national economy.

CS developments of the last decade are likely to have as profound a potential impact on productivity, labor markets, working conditions, and the quality of life in the developed countries as the introduction of robot into workplace. The conclusions can be reached based on four factors:

- First, the estimate of the number of jobs that could be performed by is relatively small.

- Second, almost all of these workers would be spared forced unemployment because of retraining and in some cases the job attrition that occurs through normal retirement.

- Third, total employment is a function of real economic growth: robots can have a positive effect on real economic growth and, therefore, a positive effect on total employment.

- Fourth, in 10 years, retraining programs can adequately shift displaced workers to new careers. In fact, the main challenge posed to policymakers by increased use of robots is not unemployment but need for retraining.

History shows that labor-saving techniques have led to improved living standards, higher real wages, and employment growth. In large measure, the robotics revolution is merely a continuation of a centuries-long trend that has resulted in enormous material progress. Protection from job loss can come through retraining programs. Working condition and job safety will improve as robots take over dangerous and undesirable forms of works.

Technological advances in computers and microprocessors are increasing the sophistication of robots, giving them some "thinking" capacity that increase potential uses. The key to usage in such area as office work depends in large part on the ability to develop "intelligent" robots capable of performing tasks that vary somewhat over time. Some industry observers believe breakthroughs may allow for extensive introduction of robotics in non-manufacturing tasks within a few years.

Three important dimension of the growth of robotics are subjects to economic analysis:

The first is the determinants of the magnitude of the growth of the robotics industry.

The second is the impact robotics unemployment.

The third is the impact that robots will have on wages, profits and prices.

There are two reasons for the growth in the use of robots, one related primarily to supply and the second primarily to demand. In the long run, robots will be increasingly utilized because the cost of traditional labor-intensive techniques is rising over time, while the cost of the capital-intensive robotic techniques is falling relative to prices generally. These costs decline because the technological advances in robotics lower the capital costs of robots per unit output.

On the other hand, some government policies may speed robotic introduction. For example, where environmental regulations lower worker productivity or raise capital costs associated with the traditional CS, the traditional technique cost line will shift upward, advancing the date at which robotic adoption becomes profitable.

Important changes in the composition of the work force have occurred over the past four decades and, in some opinions, even more massive changes lie ahead as many thousands of low-skill jobs are eliminated while at the same time large numbers of new jobs are created to meet the demands of technological advance. If serious employment displacement effects are to be avoided, development of broad-scale training programs in which the private sector plays a key role, in concert with various governmental bodies will be required.

Control System (CS) Theory

CS theory is needed for obtaining the desired motion or force needed; sensors for vision and computers for programming these devices to accomplish their desired tasks. In a nutshell, CS theory, sometimes called automation, cybernetics or systems theory is a branch of applied mathematics that deals with the design of machinery and other engineering systems so that these systems work, and work better than before.

As an example, consider the problem of controlling the temperature in a cold lecture hall. This is a standard engineering problem familiar to us all. The thermal system consists of the furnace as the heating source, and the room thermometer as the record of the temperature of the hall. The external environment we assume fixed and not belonging to the thermodynamic system under analysis. The basic heating source is the furnace, but the control of the furnace is through a thermostat, the thermostat device usually contains a thermometer to measure the current room temperature and a dial on which we set the desired room temperature. The control aspect of thermostat is that it compares the actual and the desired temperatures at each moment and then it sends an electric signal or control command to the furnace to turn the fire intensity up or down. In this case, the job of the CS engineer is to invent or design an effective thermostat.

CS theory is a Teleological Science

First consider the philosophical position of the discipline of CS theory. Within the framework of metaphysics, CS theory is a teleological science. That is, the concepts of CS involve ideas such as purpose, goal-seeking and ideal or desirable norms. These are terms of nineteenth century biology and psychology, terms of evolution will and motivation such as were introduced by Aristotle to explain the foundations of physics, but then carefully exorcized by Newton when he constructed a human geometric mechanics. So CS theory represents a synthesis of the philosophies of Aristotle and Newton showing that inanimate deterministic mechanisms can function as purposeful self-regulating organisms.

CS is an Information Science

Another philosophical aspect of CS theory is that it avoids the concepts of energy but, instead, deals with the phenomenon of information in physical systems. If we compare the furnace with the thermostat we note a great disparity of size and weight. The powerful furnace supplies quantities of energy: a concept of classical physics. Thus CS theory rests on a new category of physical reality, namely information, which is distinct from energy or matter. Possibly, this affords a new approach to the conundrum of mind versus matter, concerning which the philosophical remarked.

But what are the problems, methods and results of CS theory as they are interpreted in modern mathematical physics or engineering? In this sense CS theory deals with the inverse problem of dynamical systems. That is, suppose we have a dynamical system, for example many vibrating masses interconnected by elastic springs. Such a dynamical system is described mathematically by an array of ordinary differential equations that predict the evolution of the vibrations according to Newton's laws of motion.

Measurement of Performance

Although such measurement is not always practicable, the measurement of performance against standards should ideally be done on a forward-looking basis so that deviations may be detected in advance of their occurrence and avoided by appropriate actions. The alert, forward-looking manager can sometimes predict probable departures form standards. In the absence of such ability, however, deviations should be disclosed as early as possible.

If standards are appropriately drawn and if means are available for determining exactly what subordinates are doing, appraisal of actual or expected performance is fairly easy. But there are many activities for which it is difficult to develop accurate standards, and there are many activities that are hard to measure. It may be quite simple to establish labor-hour standards for the production of a mass-produced item, and it may be equally simple to measure performance against these standards, but if the item is customs-made, the appraisal of performance may be a formidable task because standards are difficult to set.

Correction of Deviations

Standards should reflect the various positions in an organization structure. If performance is measured accordingly, it is easier to correct deviations. Managers know exactly where, in the assignment of individual or group duties, the corrective measure must be applied.

Correction of deviations is the point at which control can be seen as a part of the whole system of management and can be related to the other managerial functions. Managers may correct deviations by redrawing their plans or by modifying their goals. This is an exercise of the principle of navigational change or they may correct deviations by exercising their organizing function through reassignment or clarification of duties. They may correct, also, by that ultimate re-staffing measure-firing or, again, they may correct through better leading-fuller explanation of the job or more effective leadership techniques.

References

- Control-engineering-historical-review-and-types-of-control-engineering: electrical4u.com, Retrieved 14 February, 2019

- Control-systems-block-diagrams, control-systems: tutorialspoint.com, Retrieved 16 June, 2019

- Control-system-block-diagram: javatpoint.com, Retrieved 25 July, 2019

- Control-systems-signal-flow-graphs, control-systems: tutorialspoint.com, Retrieved 05 May, 2019

- Digital-data-control-system: electrical4u.com, Retrieved 13 April, 2019

- Time-domain-analysis-of-control-system: electrical4u.com, Retrieved 17 January, 2019

Control Theory

The subfield of mathematics which deals with the control of continuously operating dynamical systems in engineered processes and machines is called control theory. Various methods and branches of control theory include digital control, hybrid control, multivariable control, nonlinear control, robust control, stochastic control, etc. All these methods and branches of control theory have been carefully analyzed in this chapter.

In engineering and mathematics, control theory deals with the behaviour of dynamical systems. The desired output of a system is called the reference. When one or more output variables of a system need to follow a certain reference over time, a controller manipulates the inputs to a system to obtain the desired effect on the output of the system. Rapid advances in digital system technology have radically altered the control design options. It has become routinely practicable to design very complicated digital controllers and to carry out the extensive calculations required for their design. These advances in implementation and design capability can be obtained at low cost because of the widespread availability of inexpensive and powerful digital processing platforms and high-speed analog IO devices.

ADAPTIVE CONTROL

Adaptive control is a method whereby the gain of a system can be varied depending on the position of the set point. The following shows a simple example of why this is useful in control systems.

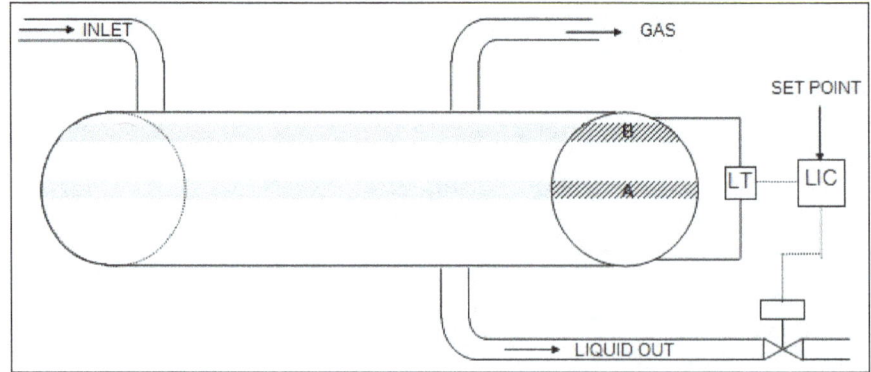

Level Control of a separator.

Shows the level control of a separator. The level in the separator can be set to control at position A or position B.

As the level changes the volume of liquid to be removed or added at position A is much greater than what must be removed or added at position B. So, for good response the gain at position A should be greater than the gain at position B. The LIC is mP based. The gain of the controller is programmed by the engineer so that it changes when the set point is changed.

Adaptive control is based on the development of a compensating adjustment through knowledge of a disturbing factor or on the basis of loop response itself. A system that adapts itself on the basis of a measurement of a disturbing factor is referred to as programmed system.

A system that uses a measurement of its own performance is termed *self-adaptive.*

On the other hand, self-adaptive systems must sense variations in plant behaviour. This is what identification is and it is normally a partial measurement e- cause total identification is either too complex or impossible. After identification, the adaptive device or system must be able to adjust some controller parameter or function. This is referred to as actuation and is illustrated in the schematic diagram of figure below.

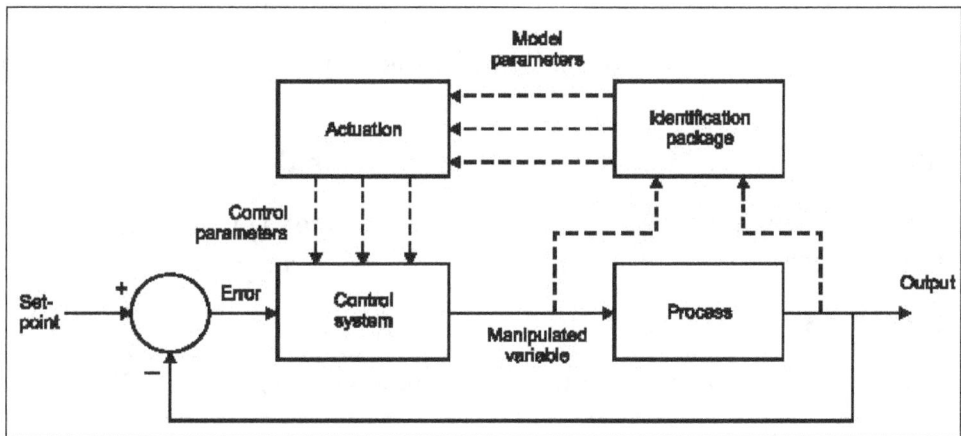

In a real sense, programmed adaptive control is feedforward in nature, while self adaptive is feedback.

Advantages

- Increased production rates.
- Increased tool life.
- Highest metal removal rates consistent to the existing conditions.
- It is more advantageous where there are wide variations in the depth of cut during machining.
- Dimensional accuracy and better surface finish can be obtained.

Disadvantages

- Unavailability of suitable sensors that have a reliable operation ina manufacturing environment.
- Force and torque sensors in systems are difficult to install.

DIGITAL CONTROL

Digital control is a branch of control theory that uses digital computers to act as system controllers. Depending on the requirements, a digital control system can take the form of a microcontroller to an ASIC to a standard desktop computer. Since a digital computer is a discrete system, the Laplace transform is replaced with the Z-transform. Also since a digital computer has finite precision (*See quantization*), extra care is needed to ensure the error in coefficients, A/D conversion, D/A conversion, etc. are not producing undesired or unplanned effects.

The application of digital control can readily be understood in the use of feedback. Since the creation of the first digital computer in the early 1940s the price of digital computers has dropped considerably, which has made them key pieces to control systems for several reasons:

- Inexpensive: Under $5 for many microcontrollers.

- Flexible: Easy to configure and reconfigure through software.

- Scalable: Programs can scale to the limits of the memory or storage space without extra cost.

- Adaptable: Parameters of the program can change with time.

- Static operation: Digital computers are much less prone to environmental conditions than capacitors, inductors, etc.

Digital Controller Implementation

A digital controller is usually cascaded with the plant in a feedback system. The rest of the system can either be digital or analog.

Typically, a digital controller requires:

- A/D conversion to convert analog inputs to machine readable (digital) format.

- D/A conversion to convert digital outputs to a form that can be input to a plant (analog).

- A program that relates the outputs to the inputs.

Output Program

Outputs from the digital controller are functions of current and past input samples, as well as past output samples - this can be implemented by storing relevant values of input and output in registers. The output can then be formed by a weighted sum of these stored values.

The programs can take numerous forms and perform many functions

- A digital filter for low-pass filtering.

- A state space model of a system to act as a state observer.

- A telemetry system.

Stability

Although a controller may be stable when implemented as an analog controller, it could be unstable when implemented as a digital controller due to a large sampling interval. During sampling the aliasing modifies the cutoff parameters. Thus the sample rate characterizes the transient response and stability of the compensated system, and must update the values at the controller input often enough so as to not cause instability.

When substituting the frequency into the z operator, regular stability criteria still apply to discrete control systems. Nyquist criteria apply to z-domain transfer functions as well as being general for complex valued functions. Bode stability criteria apply similarly. Jury criterion determines the discrete system stability about its characteristic polynomial.

Design of Digital Controller in S-domain

The digital controller can also be designed in the s-domain (continuous). The Tustin transformation can transform the continuous compensator to the respective digital compensator. The digital compensator will achieve an output which approaches the output of its respective analog controller as the sampling interval is decreased.

$$s = \frac{2(z-1)}{T(z+1)}$$

Tustin Transformation Deduction

Tustin is the Padé(1,1) approximation of the exponential function $z = e^{sT}$:

$$z = e^{sT}$$
$$= \frac{e^{sT/2}}{e^{-sT/2}}$$
$$\approx \frac{1 + sT/2}{1 - sT/2}$$

And its inverse

$$s = \frac{1}{T}\ln(z)$$
$$= \frac{2}{T}\left[\frac{z-1}{z+1} + \frac{1}{3}\left(\frac{z-1}{z+1}\right)^3 + \frac{1}{5}\left(\frac{z-1}{z+1}\right)^5 + \frac{1}{7}\left(\frac{z-1}{z+1}\right)^7 + \cdots\right]$$
$$\approx \frac{2}{T}\frac{z-1}{z+1}$$
$$= \frac{2}{T}\frac{1-z^{-1}}{1+z^{-1}}$$

We must never forget that the digital control theory is the technique to design strategies in discrete time, (and) quantized amplitude (and) in (binary) coded form to be implemented in computer systems (microcontrollers, microprocessors) that will control the analog (continuous in time and amplitude) dynamics of analog systems.

HYBRID CONTROL

A hybrid system is a dynamical system that exhibits both continuous and discrete dynamic behavior − a system that can both *flow* (described by a differential equation) and *jump* (described by a state machine or automaton). Often, the term "hybrid dynamical system" is used, to distinguish over hybrid systems such as those that combine neural nets and fuzzy logic, or electrical and mechanical drivelines. A hybrid system has the benefit of encompassing a larger class of systems within its structure, allowing for more flexibility in modeling dynamic phenomena.

In general, the *state* of a hybrid system is defined by the values of the *continuous variables* and a discrete *mode*. The state changes either continuously, according to a *flow condition*, or discretely according to a *control graph*. Continuous flow is permitted as long as so-called *invariants* hold, while discrete transitions can occur as soon as given *jump conditions* are satisfied. Discrete transitions may be associated with *events*.

Examples:

Hybrid systems have been used to model several cyber-physical systems, including physical systems with *impact*, logic-dynamic controllers, and even Internet congestion.

Bouncing Ball

A canonical example of a hybrid system is the bouncing ball, a physical system with impact. Here, the ball (thought of as a point-mass) is dropped from an initial height and bounces off the ground, dissipating its energy with each bounce. The ball exhibits continuous dynamics between each bounce; however, as the ball impacts the ground, its velocity undergoes a discrete change modeled after an inelastic collision. A mathematical description of the bouncing ball follows. Let x_1 be the height of the ball and x_2 be the velocity of the ball. A hybrid system describing the ball is as follows:

When $x \in C = \{x_1 > 0\}$, flow is governed by $\dot{x}_1 = x_2, \dot{x}_2 = -g$, where g is the acceleration due to gravity. These equations state that when the ball is above ground, it is being drawn to the ground by gravity.

When $x \in D = \{x_1 = 0\}$, jumps are governed by $x_1^+ = x_1, x_2^+ = -\gamma x_2$, where $0 < \gamma < 1$ is a dissipation factor. This is saying that when the height of the ball is zero (it has impacted the ground), its velocity is reversed and decreased by a factor of γ. Effectively, this describes the nature of the inelastic collision.

The bouncing ball is an especially interesting hybrid system, as it exhibits Zeno behavior. Zeno behavior has a strict mathematical definition, but can be described informally as the system making an *infinite* number of jumps in a *finite* amount of time. In this example, each time the ball bounces it loses energy, making the subsequent jumps (impacts with the ground) closer and closer together in time.

It is noteworthy that the dynamical model is complete if and only if one adds the contact force between the ground and the ball. Indeed, without forces, one cannot properly define the bouncing ball and the model is, from a mechanical point of view, meaningless. The simplest contact model that represents the interactions between the ball and the ground, is the complementarity relation between the force and the distance (the gap) between the ball and the ground. This is written as $0 \leq \lambda \perp x_1 \geq 0$. Such a contact model does not incorporate magnetic forces, nor gluing effects. When the complementarity relations are in, one can continue to integrate the system after the impacts have accumulated and vanished: the equilibrium of the system is well-defined as the static equilibrium of the ball on the ground, under the action of gravity compensated by the contact force λ. One also notices from basic convex analysis that the complementarity relation can equivalently be rewritten as the inclusion into a normal cone, so that the bouncing ball dynamics is a differential inclusion into a normal cone to a convex set. See Chapters 1, 2 and 3 in Acary-Brogliato's book cited below (Springer LNACM 35, 2008).

Hybrid Systems Verification

There are approaches to automatically proving properties of hybrid systems (e.g., some of the tools mentioned below). Common techniques for proving safety of hybrid systems are computation of reachable sets, abstraction refinement, and barrier certificates.

Most verification tasks are undecidable, making general verification algorithms impossible. Instead, the tools are analyzed for their capabilities on benchmark problems. A possible theoretical characterization of this is algorithms that succeed with hybrid systems verification in all robust cases implying that many problems for hybrid systems, while undecidable, are at least quasi-decidable.

Other Modeling Approaches

Two basic hybrid system modeling approaches can be classified, an implicit and an explicit one. The explicit approach is often represented by a hybrid automaton, a hybrid program or a hybrid Petri net. The implicit approach is often represented by guarded equations to result in systems of differential algebraic equations (DAEs) where the active equations may change, for example by means of a hybrid bond graph.

As a unified simulation approach for hybrid system analysis, there is a method based on DEVS formalism in which integrators for differential equations are quantized into atomic DEVS models. These methods generate traces of system behaviors in discrete event system manner which are different from discrete time systems.

Tools

- Ariadne: A C++ library for (numerically rigorous) reachability analysis of nonlinear hybrid systems.

- C2E2: Nonlinear hybrid system verifier.

- CORA: A MATLAB Toolbox for reachability analysis of cyber-physical systems, including hybrid systems.

- Flow*: A tool for reachability analysis of nonlinear hybrid systems.

- HyCreate: A Tool for Overapproximating Reachability of Hybrid Automata.

- HyEQ: A Hybrid System Solver for Matlab.

- HyPro: A C++ library for state set representations for hybrid systems reachability analysis.

- HSolver: Verification of Hybrid Systems.

- HyTech: A Model Checker for Hybrid Systems.

- JuliaReach: A Toolbox for Set-Based Reachability.

- KeYmaera: A Hybrid Theorem Prover for Hybrid Systems.

- PHAVer: Polyhedral Hybrid Automaton Verifier.

- PowerDEVS: A general-purpose software tool for DEVS modeling and simulation oriented to the simulation of hybrid systems.

- SpaceEx: State-Space Explorer.

- S-TaLiRo: A MATLAB Toolbox for verification of Hybrid Systems with respect to Temporal Logic Specifications.

INTELLIGENT CONTROL

Intelligent control is a class of control techniques that use various artificial intelligence computing approaches like neural networks, Bayesian probability, fuzzy logic, machine learning, reinforcement learning, evolutionary computation and genetic algorithms.

Intelligent control can be divided into the following major sub-domains:

- Neural network control.

- Machine learning control.

- Reinforcement learning.

- Bayesian control.

- Fuzzy control.

- Neuro-fuzzy control.

- Expert Systems.

- Genetic control.

New control techniques are created continuously as new models of intelligent behavior are created and computational methods developed to support them.

Neural Network Controller

Neural networks have been used to solve problems in almost all spheres of science and technology. Neural network control basically involves two steps:

- System identification.

- Control.

It has been shown that a feedforward network with nonlinear, continuous and differentiable activation functions have universal approximation capability. Recurrent networks have also been used for system identification. Given, a set of input-output data pairs, system identification aims to form a mapping among these data pairs. Such a network is supposed to capture the dynamics of a system. For the control part, deep reinforcement learning has shown its ability to control complex systems.

Bayesian Controllers

Bayesian probability has produced a number of algorithms that are in common use in many advanced control systems, serving as state space estimators of some variables that are used in the controller.

The Kalman filter and the Particle filter are two examples of popular Bayesian control components. The Bayesian approach to controller design often requires an important effort in deriving the so-called system model and measurement model, which are the mathematical relationships linking the state variables to the sensor measurements available in the controlled system. In this respect, it is very closely linked to the system-theoretic approach to control design.

Fuzzy Control

A fuzzy control system is a control system based on fuzzy logic—a mathematical system that analyzes analog input values in terms of logical variables that take on continuous values between 0 and 1, in contrast to classical or digital logic, which operates on discrete values of either 1 or 0 (true or false, respectively).

Fuzzy logic is widely used in machine control. The term "fuzzy" refers to the fact that the logic involved can deal with concepts that cannot be expressed as the "true" or "false" but rather as "partially true". Although alternative approaches such as genetic algorithms and neural networks can perform just as well as fuzzy logic in many cases, fuzzy logic has the advantage that the solution to the problem can be cast in terms that human operators can understand, so that their experience can be used in the design of the controller. This makes it easier to mechanize tasks that are already successfully performed by humans.

Applications

Fuzzy systems were initially implemented in Japan.

- Interest in fuzzy systems was sparked by Seiji Yasunobu and Soji Miyamoto of Hitachi, who in 1985 provided simulations that demonstrated the feasibility of fuzzy control systems for the Sendai Subway. Their ideas were adopted, and fuzzy systems were used to control accelerating, braking, and stopping when the Namboku Line opened in 1987.

- In 1987, Takeshi Yamakawa demonstrated the use of fuzzy control, through a set of simple dedicated fuzzy logic chips, in an "inverted pendulum" experiment. This is a classic control problem, in which a vehicle tries to keep a pole mounted on its top by a hinge upright by moving back and forth. Yamakawa subsequently made the demonstration more sophisticated by mounting a wine glass containing water and even a live mouse to the top of the pendulum: the system maintained stability in both cases. Yamakawa eventually went on to organize his own fuzzy-systems research lab to help exploit his patents in the field.

- Japanese engineers subsequently developed a wide range of fuzzy systems for both industrial and consumer applications. In 1988 Japan established the Laboratory for International Fuzzy Engineering (LIFE), a cooperative arrangement between 48 companies to pursue fuzzy research. The automotive company Volkswagen was the only foreign corporate member of LIFE, dispatching a researcher for a duration of three years.

- Japanese consumer goods often incorporate fuzzy systems. Matsushita vacuum cleaners use microcontrollers running fuzzy algorithms to interrogate dust sensors and adjust suction power accordingly. Hitachi washing machines use fuzzy controllers to load-weight, fabric-mix, and dirt sensors and automatically set the wash cycle for the best use of power, water, and detergent.

- Canon developed an autofocusing camera that uses a charge-coupled device (CCD) to measure the clarity of the image in six regions of its field of view and use the information provided to determine if the image is in focus. It also tracks the rate of change of lens movement during focusing, and controls its speed to prevent overshoot. The camera's fuzzy control system uses 12 inputs: 6 to obtain the current clarity data provided by the CCD and 6 to measure the rate of change of lens movement. The output is the position of the lens. The fuzzy control system uses 13 rules and requires 1.1 kilobytes of memory.

- An industrial air conditioner designed by Mitsubishi uses 25 heating rules and 25 cooling rules. A temperature sensor provides input, with control outputs fed to an inverter, a compressor valve, and a fan motor. Compared to the previous design, the fuzzy controller heats and cools five times faster, reduces power consumption by 24%, increases temperature stability by a factor of two, and uses fewer sensors.

- Other applications investigated or implemented include: character and handwriting recognition; optical fuzzy systems; robots, including one for making Japanese flower arrangements; voice-controlled robot helicopters (hovering is a "balancing act" rather similar to the inverted pendulum problem); rehabilitation robotics to provide patient-specific solutions (e.g. to control heart rate and blood pressure); control of flow of powders in film manufacture; elevator systems; and so on.

Work on fuzzy systems is also proceeding in the United State and Europe, although on a less extensive scale than in Japan.

- The US Environmental Protection Agency has investigated fuzzy control for energy-efficient motors, and NASA has studied fuzzy control for automated space docking simulations show that a fuzzy control system can greatly reduce fuel consumption.

- Firms such as Boeing, General Motors, Allen-Bradley, Chrysler, Eaton, and Whirlpool have worked on fuzzy logic for use in low-power refrigerators, improved automotive transmissions, and energy-efficient electric motors.

- In 1995 Maytag introduced an "intelligent" dishwasher based on a fuzzy controller and a "one-stop sensing module" that combines a thermistor, for temperature measurement; a conductivity sensor, to measure detergent level from the ions present in the wash; a turbidity sensor that measures scattered and transmitted light to measure the soiling of the wash; and a magnetostrictive sensor to read spin rate. The system determines the optimum wash cycle for any load to obtain the best results with the least amount of energy, detergent, and water. It even adjusts for dried-on foods by tracking the last time the door was opened, and estimates the number of dishes by the number of times the door was opened.

Research and development is also continuing on fuzzy applications in software, as opposed to firmware, design, including fuzzy expert systems and integration of fuzzy logic with neural-network and so-called adaptive "genetic" software systems, with the ultimate goal of building "self-learning" fuzzy-control systems. These systems can be employed to control complex, nonlinear dynamic plants, for example, human body.

Fuzzy Sets

The input variables in a fuzzy control system are in general mapped by sets of membership functions similar to this, known as "fuzzy sets". The process of converting a crisp input value to a fuzzy value is called "fuzzification".

A control system may also have various types of switch, or "ON-OFF", inputs along with its analog inputs, and such switch inputs of course will always have a truth value equal to either 1 or 0, but the scheme can deal with them as simplified fuzzy functions that happen to be either one value or another.

Given "mappings" of input variables into membership functions and truth values, the microcontroller then makes decisions for what action to take, based on a set of "rules", each of the form:

```
IF brake temperature IS warm AND speed IS not very fast

THEN brake pressure IS slightly decreased.
```

In this example, the two input variables are "brake temperature" and "speed" that have values defined as fuzzy sets. The output variable, "brake pressure" is also defined by a fuzzy set that can have values like "static" or "slightly increased" or "slightly decreased" etc.

Fuzzy Control in Detail

Fuzzy controllers are very simple conceptually. They consist of an input stage, a processing stage, and an output stage. The input stage maps sensor or other inputs, such as switches, thumbwheels, and so on, to the appropriate membership functions and truth values. The processing stage invokes each appropriate rule and generates a result for each, then combines the results of the rules. Finally, the output stage converts the combined result back into a specific control output value.

The most common shape of membership functions is triangular, although trapezoidal and bell curves are also used, but the shape is generally less important than the number of curves and their placement. From three to seven curves are generally appropriate to cover the required range of an input value, or the "universe of discourse" in fuzzy jargon.

The processing stage is based on a collection of logic rules in the form of IF-THEN statements, where the IF part is called the "antecedent" and the THEN part is called the "consequent". Typical fuzzy control systems have dozens of rules.

Consider a rule for a thermostat:

```
IF (temperature is "cold") THEN turn (heater is "high")
```

This rule uses the truth value of the "temperature" input, which is some truth value of "cold", to generate a result in the fuzzy set for the "heater" output, which is some value of "high". This result is used with the results of other rules to finally generate the crisp composite output. Obviously, the greater the truth value of "cold", the higher the truth value of "high", though this does not necessarily mean that the output itself will be set to "high" since this is only one rule among many. In some cases, the membership functions can be modified by "hedges" that are equivalent to adverbs. Common hedges include "about", "near", "close to", "approximately", "very", "slightly", "too", "extremely", and "somewhat". These operations may have precise definitions, though the definitions can vary considerably between different implementations. "Very", for one example, squares membership functions; since the membership values are always less than 1, this narrows the membership function. "Extremely" cubes the values to give greater narrowing, while "somewhat" broadens the function by taking the square root.

In practice, the fuzzy rule sets usually have several antecedents that are combined using fuzzy operators, such as AND, OR, and NOT, though again the definitions tend to vary: AND, in one popular definition, simply uses the minimum weight of all the antecedents, while OR uses the maximum value. There is also a NOT operator that subtracts a membership function from 1 to give the "complementary" function.

There are several ways to define the result of a rule, but one of the most common and simplest is the "max-min" inference method, in which the output membership function is given the truth value generated by the premise.

Rules can be solved in parallel in hardware, or sequentially in software. The results of all the rules that have fired are "defuzzified" to a crisp value by one of several methods. There are dozens, in theory, each with various advantages or drawbacks.

The "centroid" method is very popular, in which the "center of mass" of the result provides the crisp value. Another approach is the "height" method, which takes the value of the biggest contributor.

The centroid method favors the rule with the output of greatest area, while the height method obviously favors the rule with the greatest output value.

The diagram below demonstrates max-min inferencing and centroid defuzzification for a system with input variables "x", "y", and "z" and an output variable "n". Note that "mu" is standard fuzzy-logic nomenclature for "truth value":

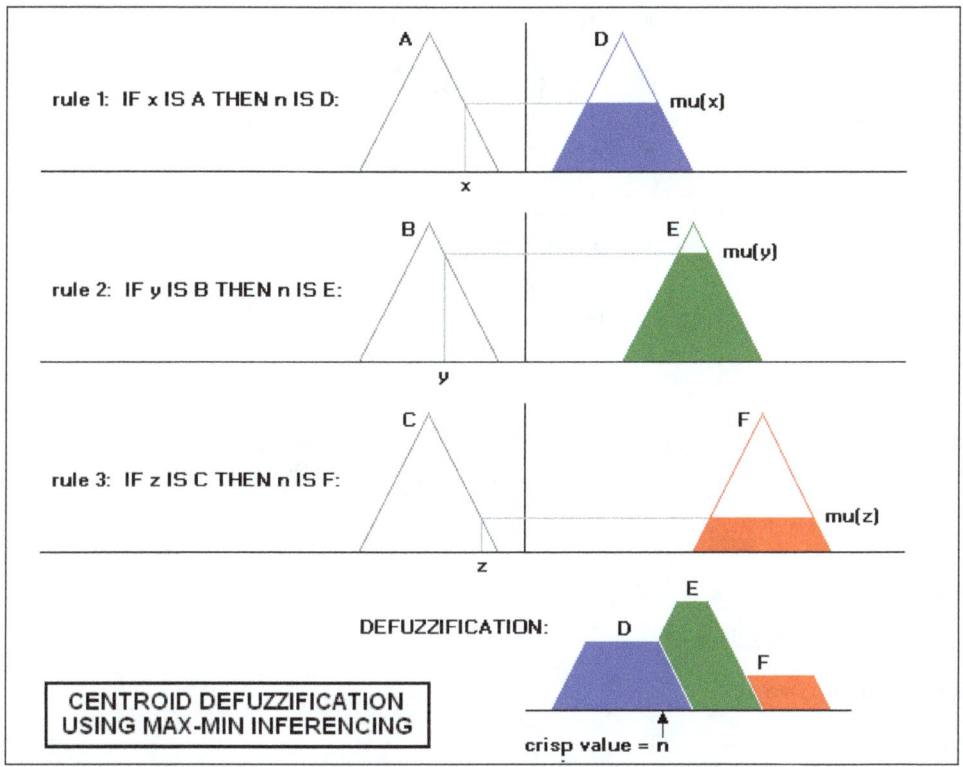

Notice how each rule provides a result as a truth value of a particular membership function for the output variable. In centroid defuzzification the values are OR'd, that is, the maximum value is used and values are not added, and the results are then combined using a centroid calculation.

Fuzzy control system design is based on empirical methods, basically a methodical approach to trial-and-error. The general process is as follows:

- Document the system's operational specifications and inputs and outputs.

- Document the fuzzy sets for the inputs.

- Document the rule set.

- Determine the defuzzification method.

- Run through test suite to validate system, adjust details as required.

- Complete document and release to production.

As a general example, consider the design of a fuzzy controller for a steam turbine. The block diagram of this control system appears as follows.

The input and output variables map into the following fuzzy set:

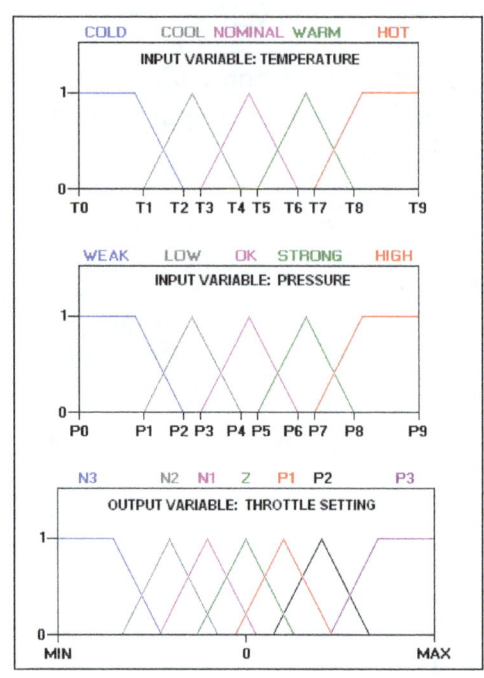

where:

N3:   Large negative.

N2:   Medium negative.

N1:   Small negative.

Z: Zero.

P1:   Small positive.

P2:   Medium positive.

P3:   Large positive.

The rule set includes such rules as:

rule 1:   IF temperature IS cool AND pressure IS weak,
          THEN throttle is P3.

rule 2:   IF temperature IS cool AND pressure IS low,
          THEN throttle is P2.

rule 3:   IF temperature IS cool AND pressure IS ok,
          THEN throttle is Z.

rule 4:   IF temperature IS cool AND pressure IS strong,
          THEN throttle is N2.

In practice, the controller accepts the inputs and maps them into their membership functions and truth values. These mappings are then fed into the rules. If the rule specifies an AND relationship between the mappings of the two input variables, as the examples above do, the minimum of the two is used as the combined truth value; if an OR is specified, the maximum is used. The appropriate output state is selected and assigned a membership value at the truth level of the premise. The truth values are then defuzzified. For an example, assume the temperature is in the "cool" state, and the pressure is in the "low" and "ok" states. The pressure values ensure that only rules 2 and 3 fire:

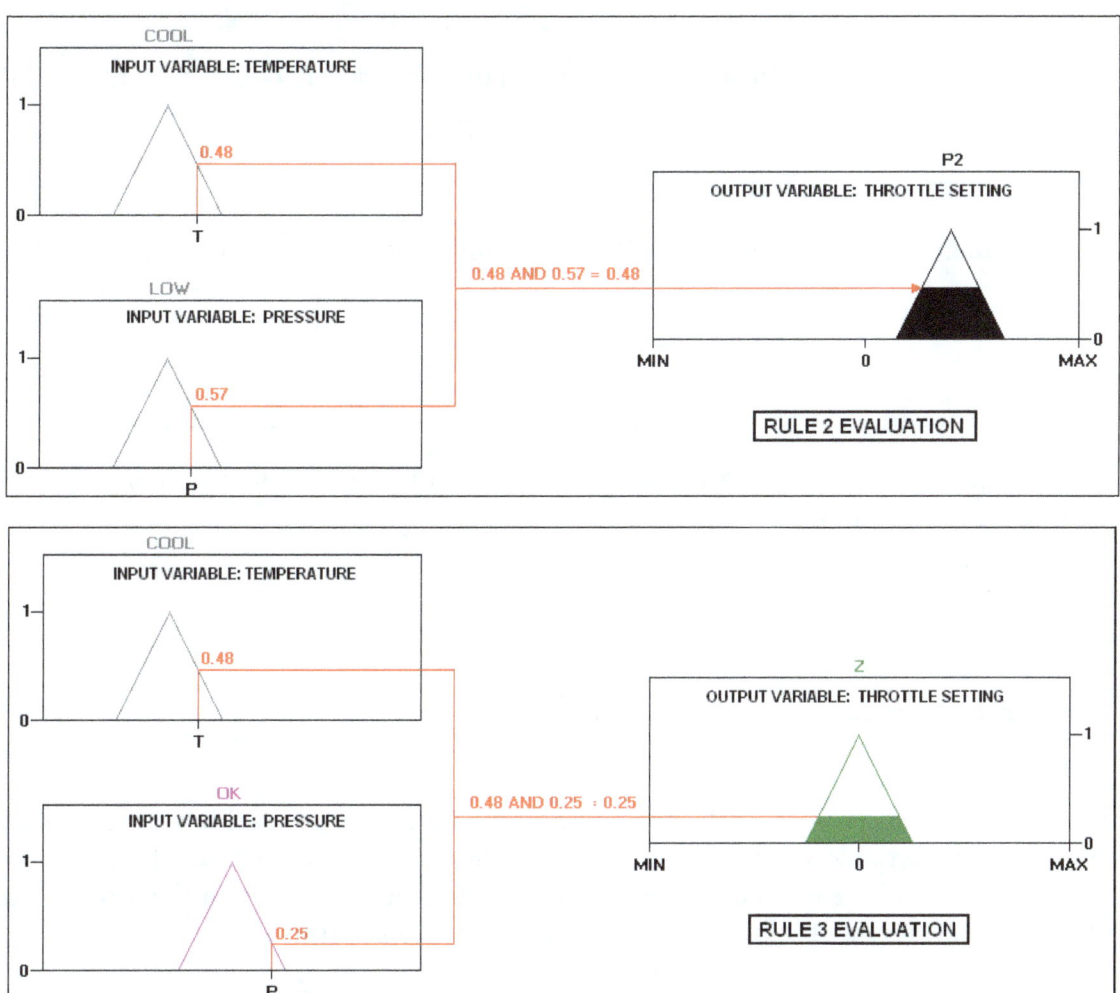

The two outputs are then defuzzified through centroid defuzzification:

```
          |              Z         P2
  1  -+              *           *
          |         *    *      *    *
          |        *      *    *       *
          |       *         *  *         *
          |     *      222222222
          |     *      22222222222
```

```
|    333333332222222222222
+---3333333322222222222222-->
              ^

            +150
```

The output value will adjust the throttle and then the control cycle will begin again to generate the next value .

Building a Fuzzy Controller

Consider implementing with a microcontroller chip a simple feedback controller:

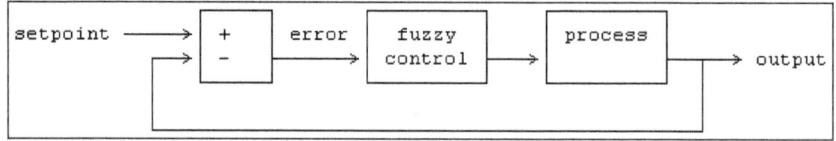

A fuzzy set is defined for the input error variable "e", and the derived change in error, "delta", as well as the "output", as follows:

```
LP:  large positive

SP:  small positive

ZE:  zero

SN:  small negative

LN:  large negative
```

If the error ranges from -1 to +1, with the analog-to-digital converter used having a resolution of 0.25, then the input variable's fuzzy set (which, in this case, also applies to the output variable) can be described very simply as a table, with the error / delta / output values in the top row and the truth values for each membership function arranged in rows beneath:

	-1	-0.75	-0.5	-0.25	0	0.25	0.5	0.7	1
mu(LP)	0	0	0	0	0	0	0.3	0.7	1
mu(SP)	0	0	0	0	0.3	0.7	1	0.7	0.3
mu(ZE)	0	0	0.3	0.7	1	0.7	0.3	0	0
mu(SN)	0.3	0.7	1	0.7	0.3	0	0	0	0
mu(LN)	1	0.7	0.3	0	0	0	0	0	0

```
—or, in graphical form (where each "X" has a value of 0.1):
      LN           SN           ZE           SP           LP
```

```
        +--------------------------------------------------------------+
        |                                                              |
 -1.0   |  XXXXXXXXXX  XXX                 :          :          :     |
 -0.75  |  XXXXXXX         XXXXXXX         :          :          :     |
 -0.5   |  XXX         XXXXXXXXXX  XXX                :          :     |
 -0.25  |  :           XXXXXXX      XXXXXXX           :          :     |
  0.0   |  :           XXX         XXXXXXXXXX  XXX               :     |
  0.25  |  :           :           XXXXXXX      XXXXXXX          :     |
  0.5   |  :           :           XXX         XXXXXXXXXX XXX          |
  0.75  |  :           :           :           XXXXXXX      XXXXXXX    |
  1.0   |  :           :           :           XXX         XXXXXXXXXX  |
        |                          |                                   |
        +--------------------------------------------------------------+
```

Suppose this fuzzy system has the following rule base:

```
        rule 1:  IF e = ZE AND delta = ZE THEN output = ZE

        rule 2:  IF e = ZE AND delta = SP THEN output = SN

        rule 3:  IF e = SN AND delta = SN THEN output = LP

        rule 4:  IF e = LP OR  delta = LP THEN output = LN
```

These rules are typical for control applications in that the antecedents consist of the logical combination of the error and error-delta signals, while the consequent is a control command output. The rule outputs can be defuzzified using a discrete centroid computation:

```
SUM( I = 1 TO 4 OF ( mu(I) * output(I) ) ) / SUM( I = 1 TO 4 OF mu(I) )
```

Now, suppose that at a given time we have:

```
        e     = 0.25

        delta = 0.5
```

Then this gives:

	e	delta
mu(LP)	0	0.3
mu(SP)	0.7	1
mu(ZE)	0.7	0.3
mu(SN)	0	0
mu(LN)	0	0

Plugging this into rule 1 gives:

```
rule 1:  IF e = ZE AND delta = ZE THEN output = ZE
    mu(1)      = MIN( 0.7, 0.3 ) = 0.3
output(1) = 0
```

-- where:

- Mu(1): Truth value of the result membership function for rule 1. In terms of a centroid calculation, this is the "mass" of this result for this discrete case.

- Output(1): Value (for rule 1) where the result membership function (ZE) is maximum over the output variable fuzzy set range. That is, in terms of a centroid calculation, the location of the "center of mass" for this individual result. This value is independent of the value of "mu". It simply identifies the location of ZE along the output range.

The other rules give:

```
rule 2:  IF e = ZE AND delta = SP THEN output = SN
    mu(2)      = MIN( 0.7, 1 ) = 0.7
    output(2) = -0.5
rule 3: IF e = SN AND delta = SN THEN output = LP
    mu(3)      = MIN( 0.0, 0.0 ) = 0
    output(3) = 1
rule 4: IF e = LP OR  delta = LP THEN output = LN
    mu(4)      = MAX( 0.0, 0.3 ) = 0.3
    output(4) = -1
```

The centroid computation yields:

$$\frac{mu(1){\cdot}output(1) + mu(2){\cdot}output(2) + mu(3){\cdot}output(3) + mu(4){\cdot}output(4)}{mu(1) + mu(2) + mu(3) + mu(4)}$$

$$= \frac{(0.3{\cdot}0) + (0.7{\cdot}{-}0.5) + (0{\cdot}1) + (0.3{\cdot}{-}1)}{0.3 + 0.7 + 0 + 0.3}$$

$= -0.5$ —for the final control output. Simple. Of course the hard part is figuring out what rules actually work correctly in practice.

If you have problems figuring out the centroid equation, remember that a centroid is defined by summing all the moments (location times mass) around the center of gravity and equating the sum to zero. So if X_0 is the center of gravity, X_i is the location of each mass, and M_i is each mass, this gives:

$$0 = (X_1 - X_0){\cdot}M_1 + (X_2 - X_0){\cdot}M_2 + \ldots + (X_n - X_0){\cdot}M_n$$

$$0 = (X_1{\cdot}M_1 + X_2{\cdot}M_2 + \ldots + X_n{\cdot}M_n) - X_0{\cdot}(M_1 + M_2 + \ldots + M_n)$$

$$X_0 \cdot (M_1 + M_2 + \ldots + M_n) = X_1 \cdot M_1 + X_2 \cdot M_2 + \ldots + X_n \cdot M_n$$

$$X_0 = \frac{X_1 \cdot M_1 + X_2 \cdot M_2 + \ldots + X_n \cdot M_n}{M_1 + M_2 + \ldots + M_n}$$

In our example, the values of mu correspond to the masses, and the values of X to location of the masses (mu, however, only 'corresponds to the masses' if the initial 'mass' of the output functions are all the same/equivalent. If they are not the same, i.e. some are narrow triangles, while others maybe wide trapizoids or shouldered triangles, then the mass or area of the output function must be known or calculated. It is this mass that is then scaled by mu and multiplied by its location X_i).

This system can be implemented on a standard microprocessor, but dedicated fuzzy chips are now available. For example, Adaptive Logic INC of San Jose, California, sells a "fuzzy chip", the AL220, that can accept four analog inputs and generate four analog outputs. A block diagram of the chip is shown below:

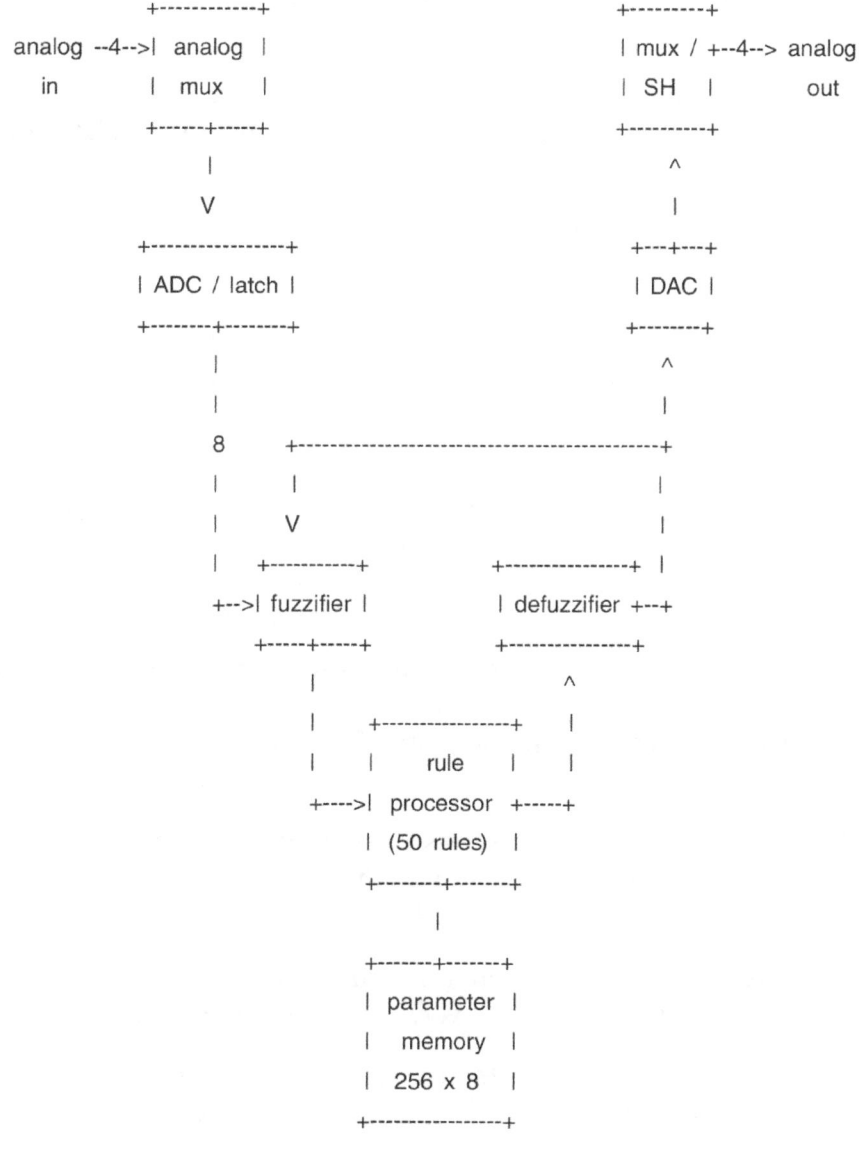

```
            +------------+                       +---------+
analog --4-->|  analog   |                       | mux / +--4--> analog
   in        |   mux     |                       |  SH  |           out
            +------+-----+                       +----------+
                   |                                  ^
                   V                                  |
          +-----------------+                     +---+---+
          | ADC / latch |                         | DAC |
          +--------+--------+                      +--------+
                   |                                  ^
                   |                                  |
                   8       +------------------------------------------+
                   |       |                          |
                   |       V                          |
                   |   +-----------+      +---------------+ |
                  +--->| fuzzifier |      | defuzzifier +---+
                      +-----+-----+        +---------------+
                            |                    ^
                            |     +---------------+   |
                            |     |     rule    |     |
                       +---->| processor +-----+
                            | (50 rules) |
                           +--------+-------+
                                    |
                           +--------+-------+
                           | parameter |
                           |  memory   |
                           |  256 x 8  |
                           +-----------------+
```

```
ADC:   analog-to-digital converter

DAC:   digital-to-analog converter

SH:    sample/hold
```

Antilock Brakes

As a first example, consider an anti-lock braking system, directed by a microcontroller chip. The microcontroller has to make decisions based on brake temperature, speed, and other variables in the system.

The variable "temperature" in this system can be subdivided into a range of "states": "cold", "cool", "moderate", "warm", "hot", "very hot". The transition from one state to the next is hard to define.

An arbitrary static threshold might be set to divide "warm" from "hot". For example, at exactly 90 degrees, warm ends and hot begins. But this would result in a discontinuous change when the input value passed over that threshold. The transition wouldn't be smooth, as would be required in braking situations.

The way around this is to make the states *fuzzy*. That is, allow them to change gradually from one state to the next. In order to do this there must be a dynamic relationship established between different factors.

We start by defining the input temperature states using "membership functions":

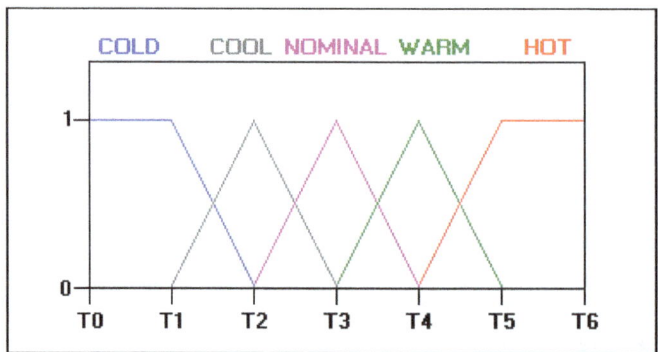

With this scheme, the input variable's state no longer jumps abruptly from one state to the next. Instead, as the temperature changes, it loses value in one membership function while gaining value in the next. In other words, its ranking in the category of cold decreases as it becomes more highly ranked in the warmer category.

At any sampled timeframe, the "truth value" of the brake temperature will almost always be in some degree part of two membership functions: i.e.: '0.6 nominal and 0.4 warm', or '0.7 nominal and 0.3 cool', and so on.

The above example demonstrates a simple application, using the abstraction of values from multiple values. This only represents one kind of data, however, in this case, temperature.

Adding additional sophistication to this braking system, could be done by additional factors such as traction, speed, inertia, set up in dynamic functions, according to the designed fuzzy system.

Neural Network Control

Multilayer neural networks have been applied successfully in the identification and control of dynamic systems. Rather than attempt to survey the many ways in which multilayer networks have been used in control systems, we will concentrate on three typical neural network controllers: model predictive control, NARMA-L2 control, and model reference control. These controllers are representative of the variety of common ways in which multilayer networks are used in control systems. As with most neural controllers, they are based on standard linear control architectures.

There are typically two steps involved when using neural networks for control: system identification and control design. In the system identification stage, we develop a neural network model of the plant that we want to control. In the control design stage, we use the neural network plant model to design (or train) the controller.

NN Predictive Control

There are a number of variations of the neural network predictive controller that are based on linear model predictive controllers. The controller then calculates the control input that will optimize plant performance over a specified future time horizon. The first step in model predictive control is to determine the neural network plant model (system identification). Next, the plant model is used by the controller to predict future performance.

System Identification

The first stage of model predictive control is to train a neural network to represent the forward dynamics of the plant. The prediction error between the plant output and the neural network output is used as the neural network training signal. The process is represented by figure.

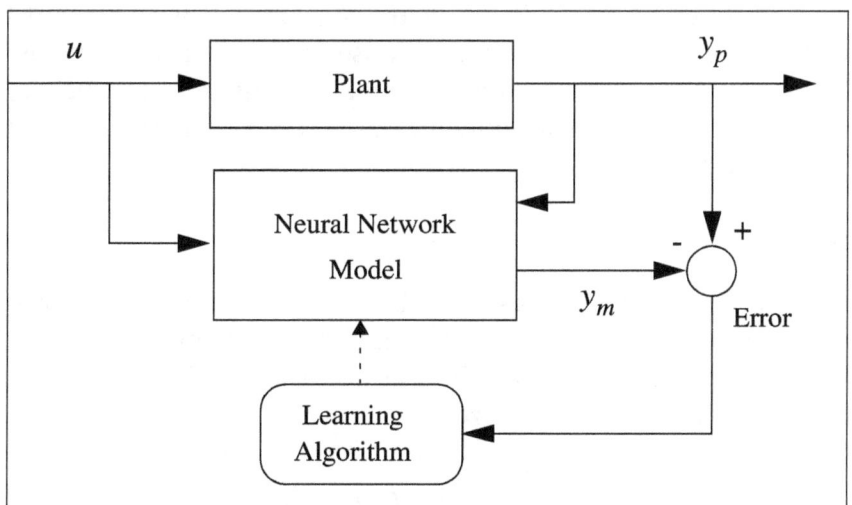

Plant Identification.

One standard model that has been used for nonlinear identification is the Nonlinear Autoregressive-Moving Average (NARMA) model:

$$y(k+d)=h\left[y(k),y(k-1),...,y(k-n+1),u(k),u(k-1),...,u(k-m+1)\right]$$

where $u(k)$ is the system input, $y(k)$ is the system output and is the system delay (we will use a delay of 1 for the predictive controller). For the identification phase, we train a neural network to approximate the nonlinear function . The structure of the neural network plant model is given in figure, where the blocks labeled TDL are tapped delay lines that store previous values of the input signal. The equation for the plant model is given by

$$y_m(k+1) = \hat{h}\big[y_p(k),...,y_p(k-n+1),u(k),...,u(k-m+1);\mathrm{x}\big],$$

where $\hat{h}[.,\mathrm{x}]$ is the function implemented by the neural network, and x is the vector containing all network weights and biases.

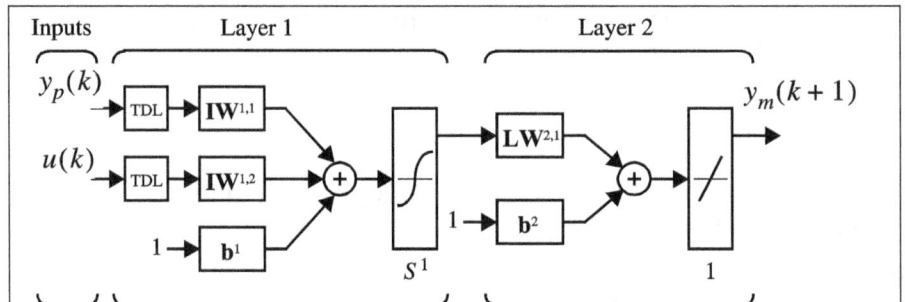

Neural Network Plant Model.

We have modified our previous notation here, to allow more than one input into the network. $\mathrm{IW}^{i,j}$ is a weight matrix from input number j to layer number i. $\mathrm{LW}^{i,j}$ is a weight matrix from layer number j to layer number i.

Although there are delays in this network, they occur only at the network input, and the network contains no feedback loops. For these reasons, the neural network plant model can be trained using the backpropagation methods for feedforward networks. It is important that the training data cover the entire range of plant operation, because we know from previous discussions that nonlinear neural networks do not extrapolate accurately. The input to this network is an $(n_y + n_u)$ -dimensional vector of previous plant outputs and inputs. It is this space that must be covered adequately by the training data.

Predictive Control

The model predictive control method is based on the receding horizon technique. The neural network model predicts the plant response over a specified time horizon. The predictions are used by a numerical optimization program to determine the control signal that minimizes the following performance criterion over the specified horizon.

$$J = \sum_{j=N_1}^{N_2}\big(y_r(k+j) - y_m(k+j)\big)^2 + \rho\sum_{j=1}^{N_u}\big(u'(k+j-1) - u'(k+j-2)\big)^2$$

where N_1, N_2 and N_u define the horizons over which the tracking error and the control increments are evaluated. The u' variable is the tentative control signal, y_r is the desired response and y_m is the network model response. The value determines the contribution that the sum of the squares of the control increments has on the performance index.

The following block diagram illustrates the model predictive control process. The controller consists of the neural network plant model and the optimization block. The optimization block determines the values of u' that minimize J, and then the optimal u is input to the plant.

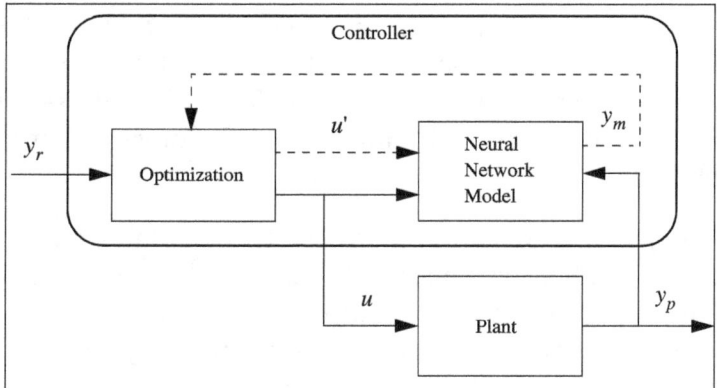

Neural Network Predictive Control.

Application - Magnetic Levitation System

Now we will demonstrate the predictive controller by applying it to a simple test problem. In this test problem, the objective is to control the position of a magnet suspended above an electromagnet, where the magnet is constrained so that it can only move in the vertical direction, as shown in figure.

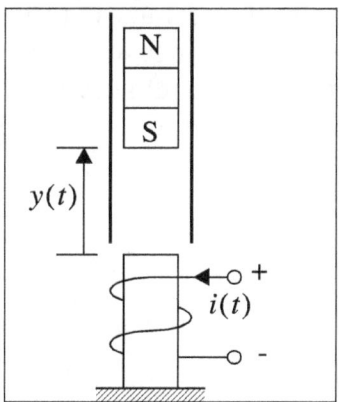

Magnetic Levitation System.

The equation of motion of the magnet is:

$$\frac{d^2 y(t)}{dt^2} = -g + \frac{\alpha}{M} \frac{i^2(t) \operatorname{sgn}[i(t)]}{y(t)} - \frac{\beta}{M} \frac{dy(t)}{dt}$$

where y(t) is the distance of the magnet above the electromagnet, i(t) is the current flowing in the electromagnet, M is the mass of the magnet, and g is the gravitational constant. The parameter β is a viscous friction coefficient that is determined by the material in which the magnet moves, and α is a field strength constant that is determined by the number of turns of wire on the electromagnet and the strength of the magnet. For our simulations, the current is allowed to range from 0 to 4 amps, and the sampling interval for the controller is 0.01 seconds. The parameter values are set to $\beta = 12$, $\alpha = 15$, $g = 9.8$ and $M = 3$.

The first step in the control design process is the development of the plant model. The performances of neural network controllers are highly dependent on the accuracy of the plant identification.

As with linear system identification, we need to insure that the plant input is sufficiently exciting. For nonlinear black box identification, we also need to be sure that the system inputs and outputs cover the operating range for which the controller will be applied. For our applications, we typically collect training data while applying random inputs which consist of a series of pulses of random amplitude and duration. The duration and amplitude of the pulses must be chosen carefully to produce accurate identification.

If the identification is poor, then the resulting control system may fail. Controller performance tends to fail in either steady state operation, or transient operation, or both. When steady state performance is poor, it is useful to increase the duration of the input pulses. Unfortunately, within a training data set, if we have too much data in steady state conditions, the training data may not be representative of typical plant behavior. This is due to the fact that the input and output signals do not adequately cover the region that is going to be controlled. This will result in poor transient performance. We need to choose the training data so that we produce adequate transient and steady state performance. The following example will illustrate the performances of the predictive controllers when we use different ranges for the pulse widths of the input signal to generate the training data.

We found that it took about 4.5 seconds for the magnetic levitation system to reach steady state in open-loop conditions. There- fore, we first specified a pulse width range of $0.01 < \tau < 5$. The neural network plant model used three delayed values of current ($m = 3$) and three delayed values of magnet position ($n = 3$) as input to the network, and 10 neurons were used in the hidden layer. After training the network with the data set shown in figure, the resulting neural network predictive control system was unstable. (The network was trained with Bayesian regularization to prevent overfitting.)

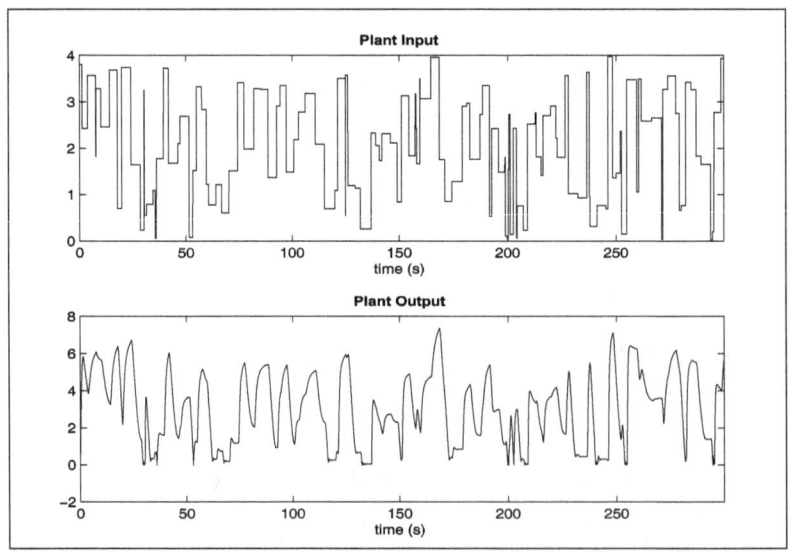

Training data with a long pulse width.

Based on the poor response of the initial controller, we determined that the training data did not provide significant coverage. Therefore, we changed the range of the input pulse widths to $0.01 < \tau < 1$, as shown in figure. From this figure, we can see that the training data is more dense and provides wider coverage of the plant model input space than the first data set. After training the network using this data set, the resulting predictive control system was stable, although it resulted in large steady- state errors.

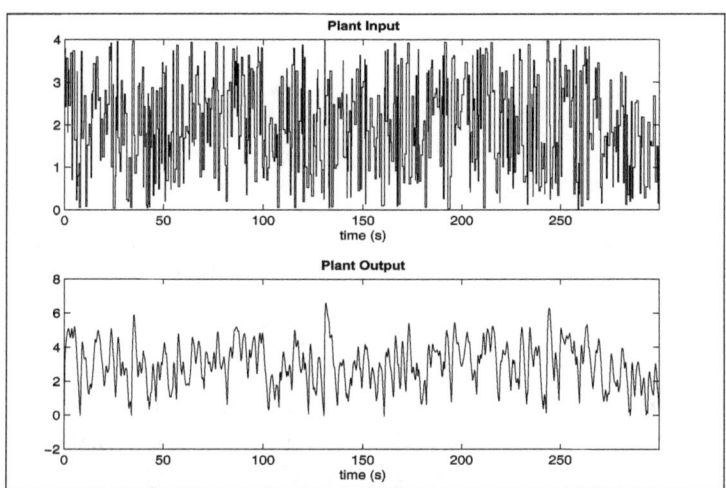

Training data with a short pulse width.

In the third test, we combined short pulse width and long pulse width (steady state) data. The long pulses were concentrated only on some ranges of plant outputs. For example, we chose to have steady-state data over the ranges where the tracking errors from the previous case were large. The input and output signals. The resulting controller performed well in both transient and steady state conditions.

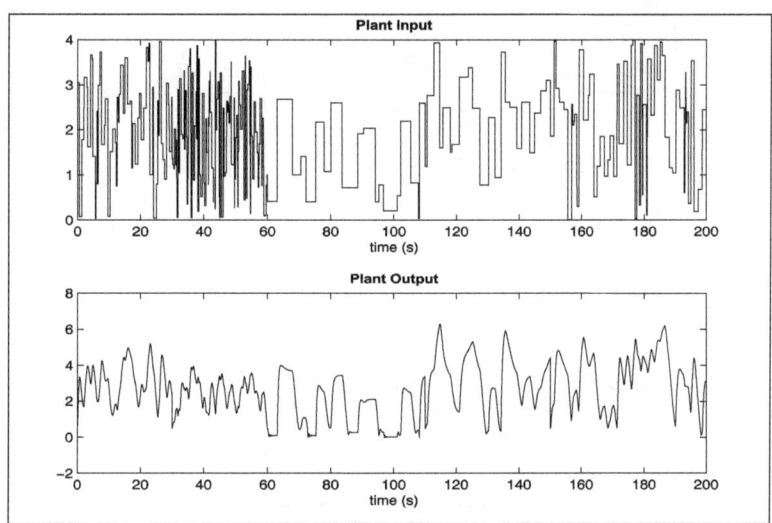

Training data with mixed pulse width.

The graph above shows the reference signal and the position of the magnet for the final predictive controller (using the neural network trained with the data shown in figure and controller parameters set to $N2 = 15$, $Nu = 3$, $\rho = 0.01$). Steady state errors were small, and the transient performance was adequate in all tested regions. We found that stability was strongly influenced by the selection of ρ. As we decrease ρ , the control signal tends to change more abruptly, generating a noisy plant output. As with linear predictive control, when we increase ρ too much, the control action is excessively smooth and the response is slow. The control action for this example is shown in the right graph of figure.

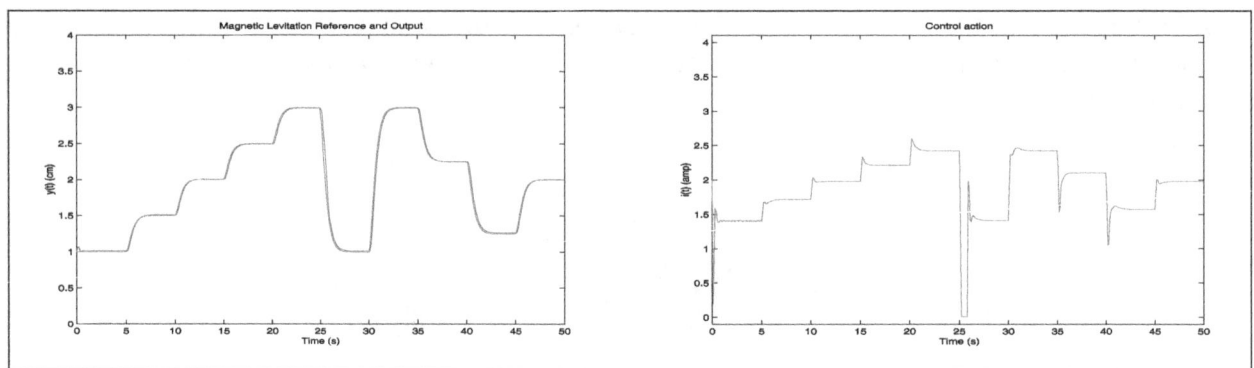

MagLev response and control action using the Predictive Controller.

NARMA-L2 Control

The neurocontroller described here is referred to by two different names: feedback linearization control and NARMA- L2 control. It is referred to as feedback linearization when the plant model has a particular form (companion form). It is referred to as NARMA-L2 control when the plant model can be approximated by the same form. The central idea of this type of control is to transform nonlinear system dynamics into linear dynamics by canceling the nonlinearities.

Identification of the NARMA-L2 Model

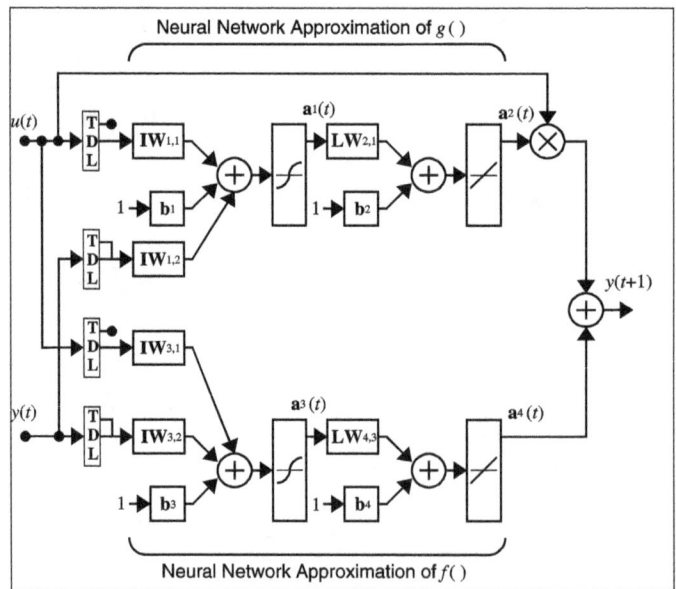

NARMA-L2 Plant Model.

As with model predictive control, the first step in using NARMA-L2 control is to identify the system to be controlled. The NARMA-L2 model is an approximation of the NARMA model of equation. The NARMA-L2 approximate model is given by:

$$\hat{y}(k+d) = f\left[y(k), y(k-1),...,y(k-n+1), u(k-1),...,u(k-m+1)\right]$$
$$+ g\left[y(k), y(k-1),...,y(k-n+1), u(k-1),...,u(k-m+1)\right] . u(k)$$

This model is in companion form, where the next controller input $u(k)$ is not contained inside the nonlinearity. Figure shows the structure of a neural network NARMA-L2 representation for $d = 1$. Notice that we have separate subnetworks to represent the functions $g(\ )$ and $f(\ )$.

NARMA-L2 Controller

The advantage of the NARMA-L2 form is that you can solve for the control input that causes the system output to follow a reference signal: $y(k+d) = y_r(k+d)$. The resulting controller would have the form

$$u(k) = \frac{y_r(k+d) - f\left[y(k), y(k-1), \ldots, y(k-n+1), u(k-1), \ldots, u(k-m+1)\right]}{g\left[y(k), y(k-1), \ldots, y(k-n+1), u(k-1), \ldots, u(k-m+1)\right]}$$

which is realizable for $d \geq 1$. Figure is a block diagram of the NARMA-L2 controller.

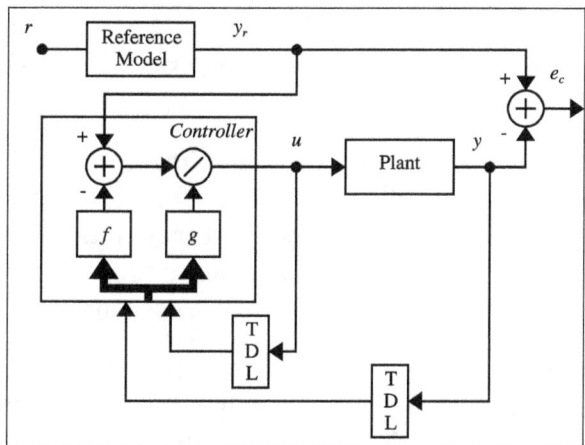

NARMA-L2 Controller.

This controller can be implemented with the previously identified NARMA-L2 plant model, as shown in figure.

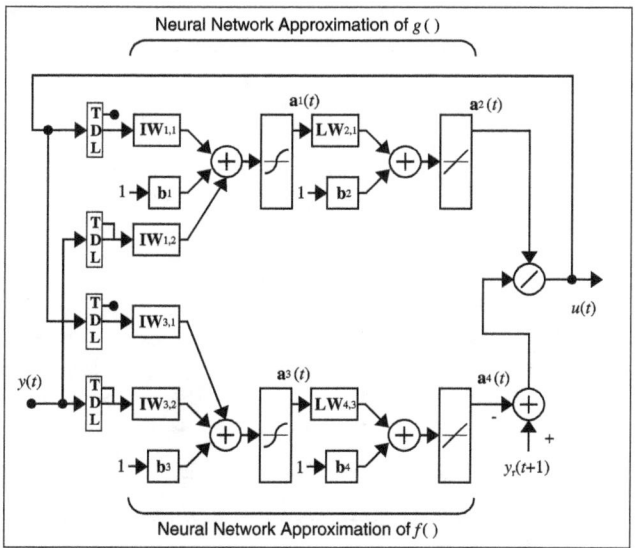

Implementation of NARMA-L2 Controller.

Application - Continuous Stirred Tank Reactor

To demonstrate the NARMA-L2 controller, we use a catalytic Continuous Stirred Tank Reactor (CSTR). The dynamic model of the system is:

$$\frac{dh(t)}{dt} = w_1(t) + w_2(t) - 0.2\sqrt{h(t)}$$

$$\frac{dC_b(t)}{dt} = \left(C_{b1} - C_b(t)\right)\frac{w_1(t)}{h(t)} + \left(C_{b2} - C_b(t)\right)\frac{w_2(t)}{h(t)} - \frac{k_1 C_b(t)}{\left(1 + k_2 C_b(t)\right)^2}$$

where $h(t)$ is the liquid level, $C_b(t)$ is the product concentration at the output of the process, $w_1(t)$ is the flow rate of the concentrated feed C_{b1}, and $w_2(t)$ is the flow rate of the diluted feed C_{b2}. The input concentrations are set to $C_{b1} = 24.9 mol/cm^3$ and $C_{b2} = 0.1 mol/cm^3$. The constants associated with the rate of consumption are $k_1 = 1$ and $k_2 = 1$. The objective of the controller is to maintain the product concentration by adjusting the flow $w_2(t)$. To simplify the demonstration, we set $w_1(t) = 0.1 cm^3/s$. The allowable range for $w_2(t)$ was assigned to be (0,4). The level of the tank $h(t)$ is not controlled for this experiment.

For the system identification phase, we used the same form of input signal as was used for the MagLev system. The pulse widths were allowed to range from 5 to 20 seconds, and the amplitude was varied from 0 to 4 cm^3/s. The neural network plant model used three delayed values of $w_2(t)$ ($n_u = 3$) and three delayed values of $C_b(t)$ ($n_y = 3$) as input to the network, and 3 neurons were used in the hidden layers. The sampling interval was set to 0.01 seconds.

The left graph of figure shows the reference signal and the system response for the NARMA-L2 controller. The output tracks the reference accurately, without significant overshoot. However, the NARMA-L2 controller generally produces more oscillatory control signals than the other controllers discussed here. This is illustrated in the control action plot shown in the right graph of figure. This chattering can be reduced by filtering (as in sliding mode control), but it is typical of NARMA-L2 controllers.

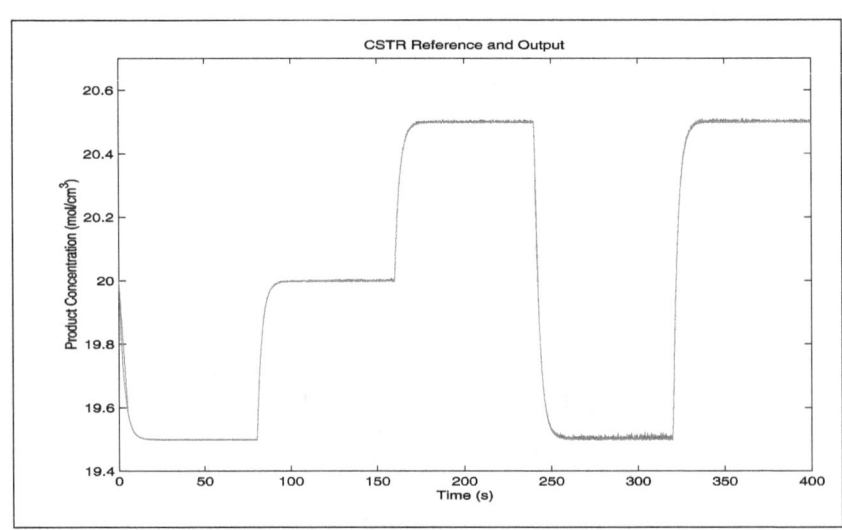

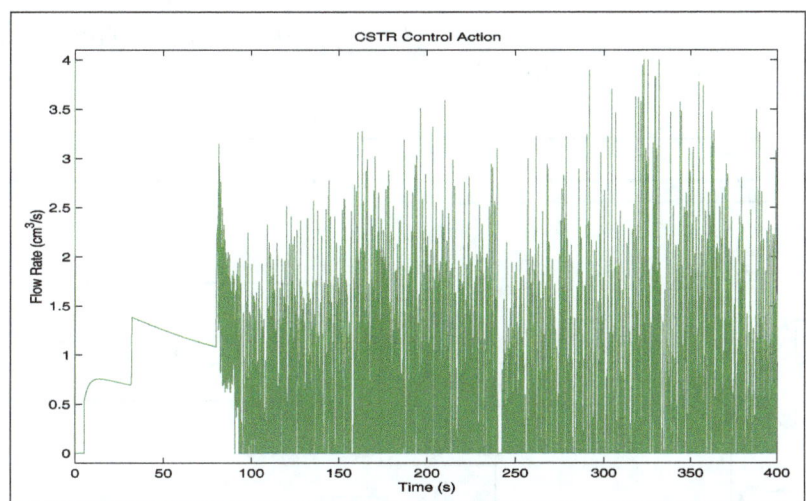

CSTR response and control action using the NARMA-L2 Controller.

Model Reference Control

This architecture uses two neural networks: a controller network and a plant model network, as shown in figure. The plant model is identified first, and then the controller is trained so that the plant output follows the reference model output.

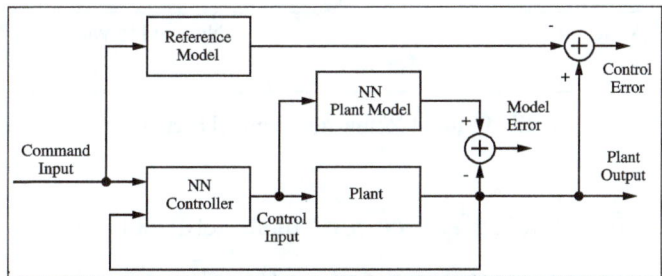

Model Reference Control Architecture.

The online computation of the model reference controller, as with NARMA-L2, is minimal. However, unlike NARMA-L2, the model reference architecture requires that a separate neural network controller be trained, in addition to the neural network plant model. The controller training is computationally expensive, since it requires the use of dynamic backpropagation. On the positive side, model reference control applies to a larger class of plant than does NARMA-L2 control, which requires that the plant be approximated by a companion form model.

Figure below shows the details of the neural network plant model and the neural network controller. There are three sets of controller inputs: Delayed reference inputs, delayed controller outputs (plant inputs), and delayed plant outputs. For each of these inputs, we select the number of delayed values to use. Typically, the number of delays increases with the order of the plant.

There are two sets of inputs to the neural network plant model: Delayed controller outputs and delayed plant outputs.

The plant identification process for model reference control is the same as that for the model predictive control, and uses the same NARMA model given by equation

$y(k+d)=h\left[y(k),y(k-1),...,y(k-n+1),u(k),u(k-1),...,u(k-m+1)\right]$. The training of the neural network controller, however, is more complex.

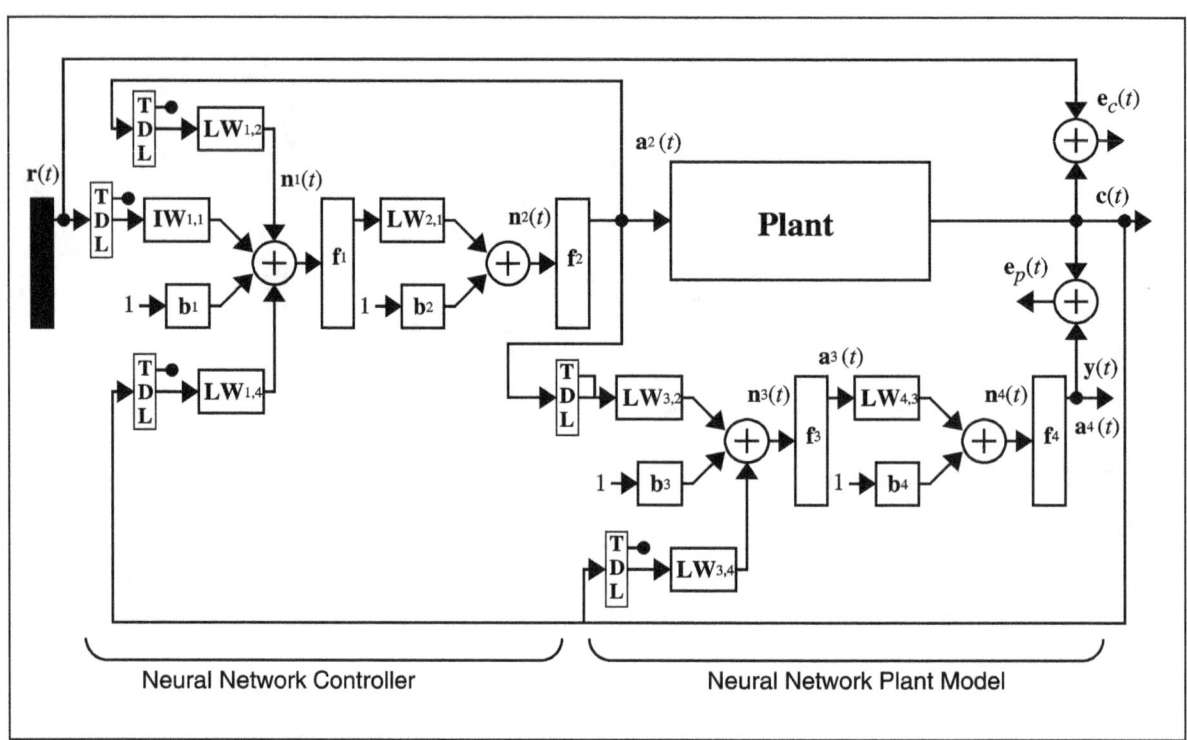

Detailed Model Reference Control Structure.

It is clear from figurethat the model reference control structure is a recurrent (feedback) network. Suppose that we use the same gradient descent algorithm, Eq. $w_{i,\ j}^{m}(k+1)=w_{i,\ j}^{m}(k)-\alpha\dfrac{\partial\hat{F}}{\partial w_{i,j}^{m}}$, , that is used in the standard backpropagation algorithm. The problem with recurrent networks is that when we try to find the equivalent of Eq. $\dfrac{\partial\hat{F}}{\partial w_{i,j}^{m}}=s_{i}^{m}a_{j}^{m-1}$ (gradient calculation) we note that the weights and biases have two different effects on the network output. The first is the direct effect, which is accounted for by Eq. $\dfrac{\partial\hat{F}}{\partial w_{i,j}^{m}}=s_{i}^{m}a_{j}^{m-1}$. The second is an indirect effect, since some of the inputs to the network, such as $u(t-1)$, are also functions of the weights and biases. To account for this indirect effect we must use dynamic backpropagation to compute the gradients for recurrent networks.

Application - Robot Arm

We will demonstrate the model reference controller on the simple, single-link robot arm shown in figure.

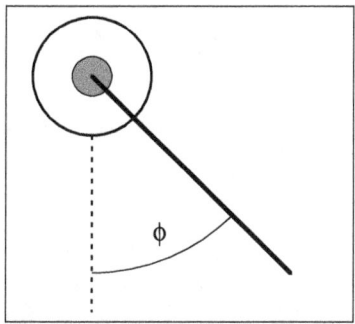

Single-link robot arm.

The equation of motion for the arm is

$$\frac{d^2\phi}{dt^2} = -10\sin\phi - 2\frac{d\phi}{dt} + u$$

where φ is the angle of the arm, and u is the torque supplied by the DC motor. The system was sampled at intervals of 0.05 seconds. To identify the system, we used input pulses with intervals between 0.1 and 2 seconds, and amplitudes between -15 and +15 N-m. The neural network plant model used two delayed values of torque ($m = 2$) and two delayed values of arm position ($n = 2$) as input to the network, and 10 neurons were used in the hidden layer (a 5-10-1 network).

The objective of the control system is to have the arm track the reference model

$$\frac{d^2 y_r}{dt^2} = -9y_r - 6\frac{dy_r}{dt} + 9r.$$

where y$_r$ is the output of the reference model, and r is the input reference signal. For the controller network, we used a 5-13- 1 architecture. The inputs to the controller consisted of two delayed reference inputs, two delayed plant outputs, and one delayed controller output. The controller was trained using a BFGS quasi-Newton algorithm, with dynamic backpropagation used to calculate the gradients.

The left graph of figure shows the reference signal and the arm position using the trained model reference controller. The system is able to follow the reference, and the control actions (shown in the right graph) are smooth. At certain set points there is some steady state error. This error could be reduced by adding more training data at those steady state conditions where the error is largest. The problem can occur in the plant model, or in the controller network.

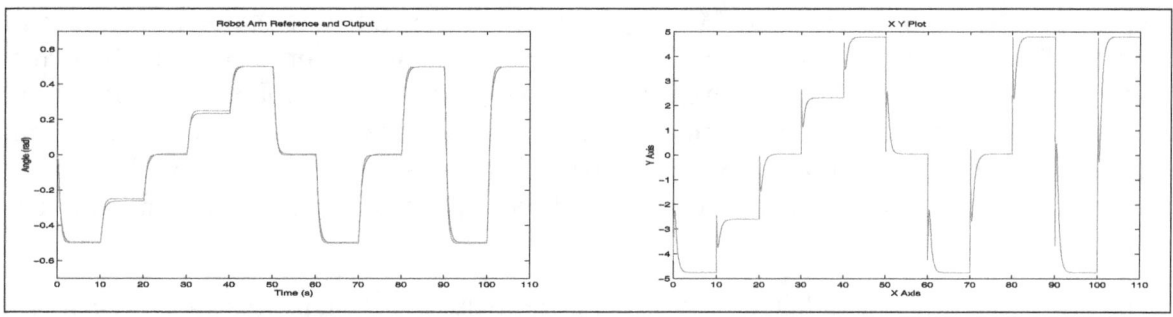

Robot arm response and control action for the Model Reference Controller.

MODEL PREDICTIVE CONTROL

Model predictive control (MPC) is an advanced method of process control that is used to control a process while satisfying a set of constraints. It has been in use in the process industries in chemical plants and oil refineries since the 1980s. In recent years it has also been used in power system balancing models and in power electronics. Model predictive controllers rely on dynamic models of the process, most often linear empirical models obtained by system identification. The main advantage of MPC is the fact that it allows the current timeslot to be optimized, while keeping future timeslots in account. This is achieved by optimizing a finite time-horizon, but only implementing the current timeslot and then optimizing again, repeatedly, thus differing from Linear-Quadratic Regulator (LQR). Also MPC has the ability to anticipate future events and can take control actions accordingly. PID controllers do not have this predictive ability. MPC is nearly universally implemented as a digital control, although there is research into achieving faster response times with specially designed analog circuitry.

Generalized predictive control (GPC) and dynamic matrix control (DMC) are classical examples of MPC.

The models used in MPC are generally intended to represent the behavior of complex dynamical systems. The additional complexity of the MPC control algorithm is not generally needed to provide adequate control of simple systems, which are often controlled well by generic PID controllers. Common dynamic characteristics that are difficult for PID controllers include large time delays and high-order dynamics.

MPC models predict the change in the dependent variables of the modeled system that will be caused by changes in the independent variables. In a chemical process, independent variables that can be adjusted by the controller are often either the setpoints of regulatory PID controllers (pressure, flow, temperature, etc.) or the final control element (valves, dampers, etc.). Independent variables that cannot be adjusted by the controller are used as disturbances. Dependent variables in these processes are other measurements that represent either control objectives or process constraints.

MPC uses the current plant measurements, the current dynamic state of the process, the MPC models, and the process variable targets and limits to calculate future changes in the dependent variables. These changes are calculated to hold the dependent variables close to target while honoring constraints on both independent and dependent variables. The MPC typically sends out only the first change in each independent variable to be implemented, and repeats the calculation when the next change is required.

While many real processes are not linear, they can often be considered to be approximately linear over a small operating range. Linear MPC approaches are used in the majority of applications with the feedback mechanism of the MPC compensating for prediction errors due to structural mismatch between the model and the process. In model predictive controllers that consist only of linear models, the superposition principle of linear algebra enables the effect of changes in multiple independent variables to be added together to predict the response of the dependent variables. This simplifies the control problem to a series of direct matrix algebra calculations that are fast and robust.

When linear models are not sufficiently accurate to represent the real process nonlinearities, several approaches can be used. In some cases, the process variables can be transformed before and after the linear MPC model to reduce the nonlinearity. The process can be controlled with nonlinear MPC that uses a nonlinear model directly in the control application. The nonlinear model may be in the form of an empirical data fit (e.g. artificial neural networks) or a high-fidelity dynamic model based on fundamental mass and energy balances. The nonlinear model may be linearized to derive a Kalman filter or specify a model for linear MPC.

An algorithmic study by El-Gherwi, Budman, and El Kamel shows that utilizing a dual-mode approach can provide significant reduction in online computations while maintaining comparative performance to a non-altered implementation. The proposed algorithm solves N convex optimization problems in parallel based on exchange of information among controllers.

Theory behind MPC

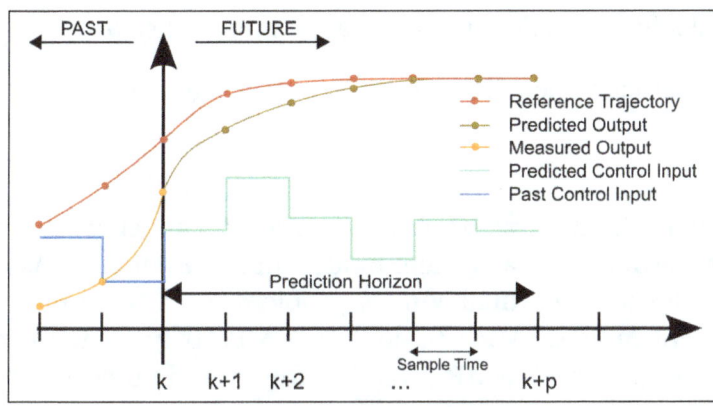

A discrete MPC scheme.

MPC is based on iterative, finite-horizon optimization of a plant model. At time t the current plant state is sampled and a cost minimizing control strategy is computed (via a numerical minimization algorithm) for a relatively short time horizon in the future: $[t, t+T]$. Specifically, an online or on-the-fly calculation is used to explore state trajectories that emanate from the current state and find (via the solution of Euler–Lagrange equations) a cost-minimizing control strategy until time $t+T$. Only the first step of the control strategy is implemented, then the plant state is sampled again and the calculations are repeated starting from the new current state, yielding a new control and new predicted state path. The prediction horizon keeps being shifted forward and for this reason MPC is also called receding horizon control. Although this approach is not optimal, in practice it has given very good results. Much academic research has been done to find fast methods of solution of Euler–Lagrange type equations, to understand the global stability properties of MPC's local optimization, and in general to improve the MPC method. To some extent the theoreticians have been trying to catch up with the control engineers when it comes to MPC.

Principles of MPC

Model Predictive Control (MPC) is a multivariable control algorithm that uses:

- An internal dynamic model of the process.
- A cost function J over the receding horizon.
- An optimization algorithm minimizing the cost function J using the control input u.

An example of a non-linear cost function for optimization is given by:

$$J = \sum_{i=1}^{N} w_{x_i} (r_i - x_i)^2 + \sum_{i=1}^{N} w_{u_i} \Delta u_i^2$$

without violating constraints (low/high limits) with

x_i : i th controlled variable (e.g. measured temperature)

r_i : i th reference variable (e.g. required temperature)

u_i : i th manipulated variable (e.g. control valve)

w_{x_i} : weighting coefficient reflecting the relative importance of x_i

w_{u_i} : weighting coefficient penalizing relative big changes in u_i etc.

Nonlinear MPC

Nonlinear Model Predictive Control, or NMPC, is a variant of model predictive control (MPC) that is characterized by the use of nonlinear system models in the prediction. As in linear MPC, NMPC requires the iterative solution of optimal control problems on a finite prediction horizon. While these problems are convex in linear MPC, in nonlinear MPC they are not necessarily convex anymore. This poses challenges for both NMPC stability theory and numerical solution.

The numerical solution of the NMPC optimal control problems is typically based on direct optimal control methods using Newton-type optimization schemes, in one of the variants: direct single shooting, direct multiple shooting methods, or direct collocation. NMPC algorithms typically exploit the fact that consecutive optimal control problems are similar to each other. This allows to initialize the Newton-type solution procedure efficiently by a suitably shifted guess from the previously computed optimal solution, saving considerable amounts of computation time. The similarity of subsequent problems is even further exploited by path following algorithms (or "real-time iterations") that never attempt to iterate any optimization problem to convergence, but instead only take a few iterations towards the solution of the most current NMPC problem, before proceeding to the next one, which is suitably initialized.

While NMPC applications have in the past been mostly used in the process and chemical industries with comparatively slow sampling rates, NMPC is being increasingly applied, with advancements in controller hardware and computational algorithms, e.g., preconditioning, to applications with high sampling rates, e.g., in the automotive industry, or even when the states are distributed in space (Distributed parameter systems). As an application in aerospace, recently, NMPC has been used to track optimal terrain-following/avoidance trajectories in real-time.

Explicit MPC

Explicit MPC (eMPC) allows fast evaluation of the control law for some systems, in stark contrast to the online MPC. Explicit MPC is based on the parametric programming technique, where the solution to the MPC control problem formulated as optimization problem is pre-computed offline. This offline solution, i.e., the control law, is often in the form of a piecewise affine function (PWA), hence the eMPC controller stores the coefficients of the PWA for each a subset (control region) of the state space, where the PWA is constant, as well as coefficients of some parametric representations of all the regions. Every region turns out to geometrically be a convex polytope for linear MPC, commonly parameterized by coefficients for its faces, requiring quantization accuracy analysis. Obtaining the optimal control action is then reduced to first determining the region containing the current state and second a mere evaluation of PWA using the PWA coefficients stored for all regions. If the total number of the regions is small, the implementation of the eMPC does not require significant computational resources (compared to the online MPC) and is uniquely suited to control systems with fast dynamics. A serious drawback of eMPC is exponential growth of the total number of the control regions with respect to some key parameters of the controlled system, e.g., the number of states, thus dramatically increasing controller memory requirements and making the first step of PWA evaluation, i.e. searching for the current control region, computationally expensive.

Robust MPC

Robust variants of Model Predictive Control (MPC) are able to account for set bounded disturbance while still ensuring state constraints are met. There are three main approaches to robust MPC:

- Min-max MPC. In this formulation, the optimization is performed with respect to all possible evolutions of the disturbance. This is the optimal solution to linear robust control problems, however it carries a high computational cost.

- Constraint Tightening MPC. Here the state constraints are enlarged by a given margin so that a trajectory can be guaranteed to be found under any evolution of disturbance.

- Tube MPC. This uses an independent nominal model of the system, and uses a feedback controller to ensure the actual state converges to the nominal state. The amount of separation required from the state constraints is determined by the robust positively invariant (RPI) set, which is the set of all possible state deviations that may be introduced by disturbance with the feedback controller.

- Multi-stage MPC. This uses a scenario-tree formulation by approximating the uncertainty space with a set of samples and the approach is non-conservative because it takes into account that the measurement information is available at every time stages in the prediction and the decisions at every stage can be different and can act as recourse to counteract the effects of uncertainties. The drawback of the approach however is that the size of the problem grows exponentially with the number of uncertainties and the prediction horizon.

MPC vs. LQR

The main differences between MPC and LQR are that LQR optimizes in a fixed time window (horizon) whereas MPC optimizes in a receding time window, and that a new solution is computed often whereas LQR uses the single (optimal) solution for the whole time horizon. Therefore, MPC typically solves the optimization problem in smaller time windows than the whole horizon and hence may obtain a suboptimal solution. However because MPC makes no assumptions about linearity, it can handle hard constraints as well as migration of a nonlinear system away from its linearized operating point, both of which are downsides of LQR.

MULTIVARIABLE CONTROL

Multivariable processes has more than one input variables or one than one output variables. Here are a few examples of multivariable processes:

- A heated liquid tank where both the level and the temperature shall be controlled.

- A distillation column where the top and bottom concentration shall be controlled.

- A robot manipulator where the positions of the manipulators (arms) shall be controlled.

- A chemical reactor where the concentration and the temperature shall be controlled.

- A head box (in a paper factory) where the bottom pressure and the paper mass level in the head box shall be controlled.

To each variable (process output variable) which is to be controlled a setpoint is given. To control these variables a number of control variables are available for manipulation by the controller function.

Multivariable processes can be difficult to control if there are cross couplings in the process, that is, if one control variable gives a response in several process output variables. There are mainly two problems of controlling a multivariable process if these cross couplings are not counteracted by the multivariable controller:

- A change in one setpoint will cause a response in each of the process output variables, not only in the output variable corresponding to the setpoint.

- Assuming that ordinary single loop PID control is used, a controller will "observe" a complicated dynamic system which consists of the multivariable process with all control loops! This can make it difficult to tune each of the PID controllers, and the stability robustness of the control system may be small.

The following section describes the most common way to control multivariable processes — namely single loop control with PID controllers.

Single loop control with PID controllers

The simplest yet most common way to control a multivariable process is using single loop control with PID controllers. There is one control loop for each process output variable which is to be controlled. The control system structure is shown in figure below, where subsystems are represented by transfer functions although these subsystems are generally non-linear dynamic systems. Since this process has two control variables and two process output variables, we say that the process is a 2x2 multivariable process.

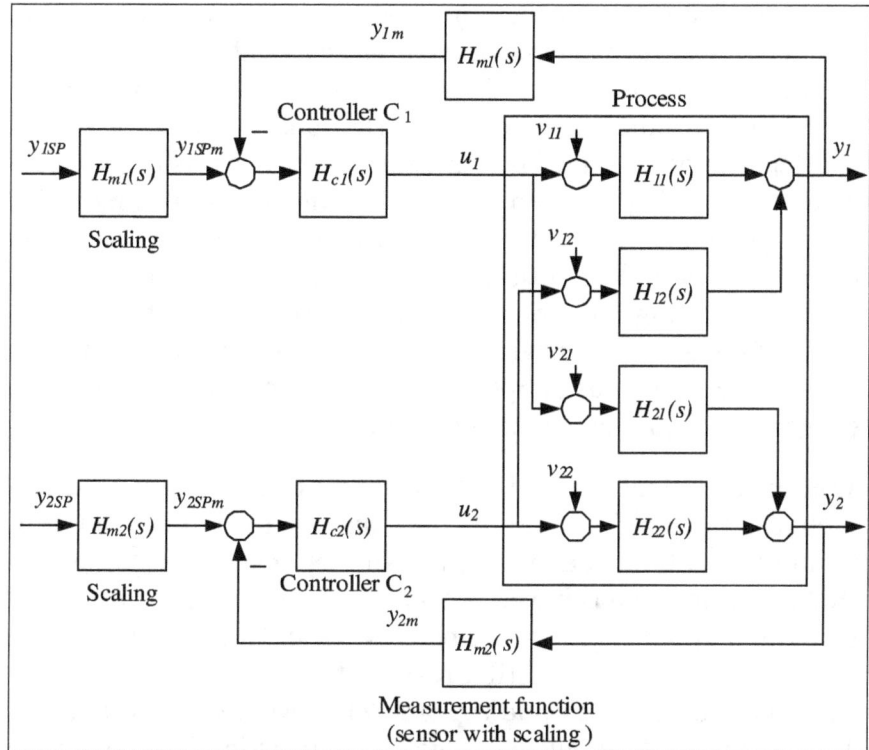

Single loop control of a 2x2 multivariable process.

NONLINEAR CONTROL

A feedback control system. It is desired to control a system (often called the *plant*) so its output follows a desired *reference* signal. A *sensor* monitors the output and a *controller* subtracts the actual output from the desired reference output, and applies this error signal to the system to bring the output closer to the reference. In a nonlinear control system at least one of the blocks, system, sensor, or controller, is nonlinear.

Nonlinear control theory is the area of control theory which deals with systems that are nonlinear, time-variant, or both. Control theory is an interdisciplinary branch of engineering and mathematics that is concerned with the behavior of dynamical systems with inputs, and how to modify the output by changes in the input using feedback, feedforward, or signal filtering. The system to be controlled is called the "plant". One way to make the output of a system follow a desired reference signal is to compare the output of the plant to the desired output, and provide feedback to the plant to modify the output to bring it closer to the desired output.

Control theory is divided into two branches. Linear control theory applies to systems made of devices which obey the superposition principle. They are governed by linear differential equations. A major subclass is systems which in addition have parameters which do not change with time, called *linear time invariant* (LTI) systems. These systems can be solved by powerful frequency domain mathematical techniques of great generality, such as the Laplace transform, Fourier transform, Z transform, Bode plot, root locus, and Nyquist stability criterion.

Nonlinear control theory covers a wider class of systems that do not obey the superposition principle. It applies to more real-world systems, because all real control systems are nonlinear. These systems are often governed by nonlinear differential equations. The mathematical techniques which have been developed to handle them are more rigorous and much less general, often applying only to narrow categories of systems. These include limit cycle theory, Poincaré maps, Lyapunov stability theory, and describing functions. If only solutions near a stable point are of interest, nonlinear systems can often be linearized by approximating them by a linear system obtained by expanding the nonlinear solution in a series, and then linear techniques can be used. Nonlinear systems are often analyzed using numerical methods on computers, for example by simulating their operation using a simulation language. Even if the plant is linear, a nonlinear controller can often have attractive features such as simpler implementation, faster speed, more accuracy, or reduced control energy, which justify the more difficult design procedure.

An example of a nonlinear control system is a thermostat-controlled heating system. A building heating system such as a furnace has a nonlinear response to changes in temperature; it is either "on" or "off", it does not have the fine control in response to temperature differences that a proportional (linear) device would have. Therefore, the furnace is off until the temperature falls below the "turn on" setpoint of the thermostat, when it turns on. Due to the heat added by the furnace, the temperature increases until it reaches the "turn off" setpoint of the thermostat, which turns the furnace off, and the cycle repeats. This cycling of the temperature about the desired temperature is called a *limit cycle*, and is characteristic of nonlinear control systems.

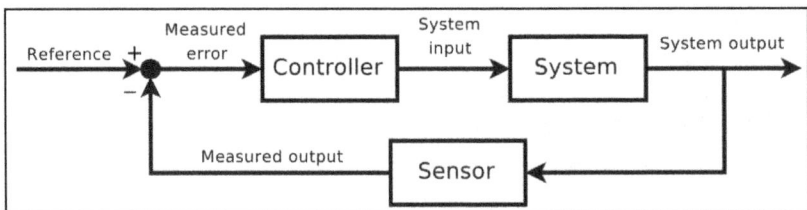

Properties of Nonlinear Systems

Some properties of nonlinear dynamic systems are

- They do not follow the principle of superposition (linearity and homogeneity).

- They may have multiple isolated equilibrium points.

- They may exhibit properties such as limit cycle, bifurcation, chaos.

- Finite escape time: Solutions of nonlinear systems may not exist for all times.

Analysis and Control of Nonlinear Systems

There are several well-developed techniques for analyzing nonlinear feedback systems:

- Describing function method.

- Phase plane method.

- Lyapunov stability analysis.

- Singular perturbation method.

- The Popov criterion and the circle criterion for absolute stability.

- Center manifold theorem.

- Small-gain theorem.

- Passivity analysis.

Control design techniques for nonlinear systems also exist. These can be subdivided into techniques which attempt to treat the system as a linear system in a limited range of operation and use (well-known) linear design techniques for each region:

- Gain scheduling.

Those that attempt to introduce auxiliary nonlinear feedback in such a way that the system can be treated as linear for purposes of control design:

- Feedback linearization.

Lyapunov based methods:

- Lyapunov redesign.

- Control-Lyapunov function.

- Nonlinear damping.

- Backstepping.

- Sliding mode control.

Nonlinear Feedback Analysis – the Lur'e Problem

An early nonlinear feedback system analysis problem was formulated by A. I. Lur'e. Control systems described by the Lur'e problem have a forward path that is linear and time-invariant, and a feedback path that contains a memory-less, possibly time-varying, static nonlinearity.

The linear part can be characterized by four matrices (A,B,C,D), while the nonlinear part is $\Phi(y)$ with $\dfrac{\Phi(y)}{y} \in [a,b], \quad a < b \quad \forall y$ (a sector nonlinearity).

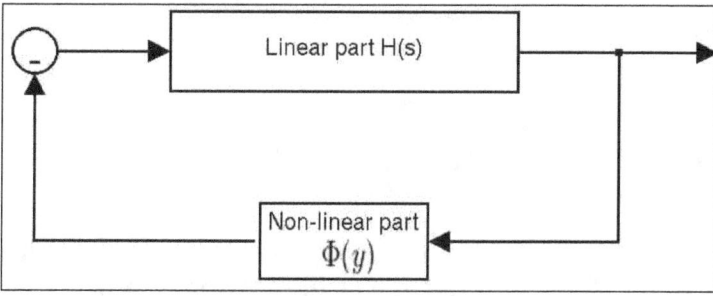

Lur'e problem block diagram.

Absolute Stability Problem

Consider:

- (A,B) is controllable and (C,A) is observable

- Two real numbers a, b with $a < b$, defining a sector for function Φ

The Lur'e problem (also known as the absolute stability problem) is to derive conditions involving only the transfer matrix $H(s)$ and $\{a,b\}$ such that $x = 0$ is a globally uniformly asymptotically stable equilibrium of the system.

There are two well-known wrong conjectures on the absolute stability problem:

- The Aizerman's conjecture

- The Kalman's conjecture.

Graphically, these conjectures can be interpreted in terms of graphical restrictions on the graph of $\Phi(y)$ x y or also on the graph of $d\Phi/dy$ x Φ/y. There are counterexamples to Aizerman's and Kalman's conjectures such that nonlinearity belongs to the sector of linear stability and unique stable equilibrium coexists with a stable periodic solution—hidden oscillation.

There are two main theorems concerning the Lur'e problem which give sufficient conditions for absolute stability:

- The circle criterion (an extension of the Nyquist stability criterion for linear systems)

- The Popov criterion.

Theoretical Results in Nonlinear Control

Frobenius Theorem

The Frobenius theorem is a deep result in differential geometry. When applied to nonlinear control, it says the following: Given a system of the form,

$$\dot{x} = \sum_{i=1}^{k} f_i(x)u_i(t)$$

where $x \in R^n$, $f_1, \ldots, f_k$ are vector fields belonging to a distribution Δ and $u_i(t)$ are control functions, the integral curves of x are restricted to a manifold of dimension m if $\text{span}(\Delta) = m$ and Δ is an involutive distribution.

REAL-TIME CONTROL

Example of a RCS-3 application of a machining workstation containing a machine tool, part buffer, and robot with vision system. RCS-3 produces a layered graph of processing nodes, each of which contains a task decomposition (TD), world modeling (WM), and sensory processing (SP) module. These modules are richly interconnected to each other by a communications system.

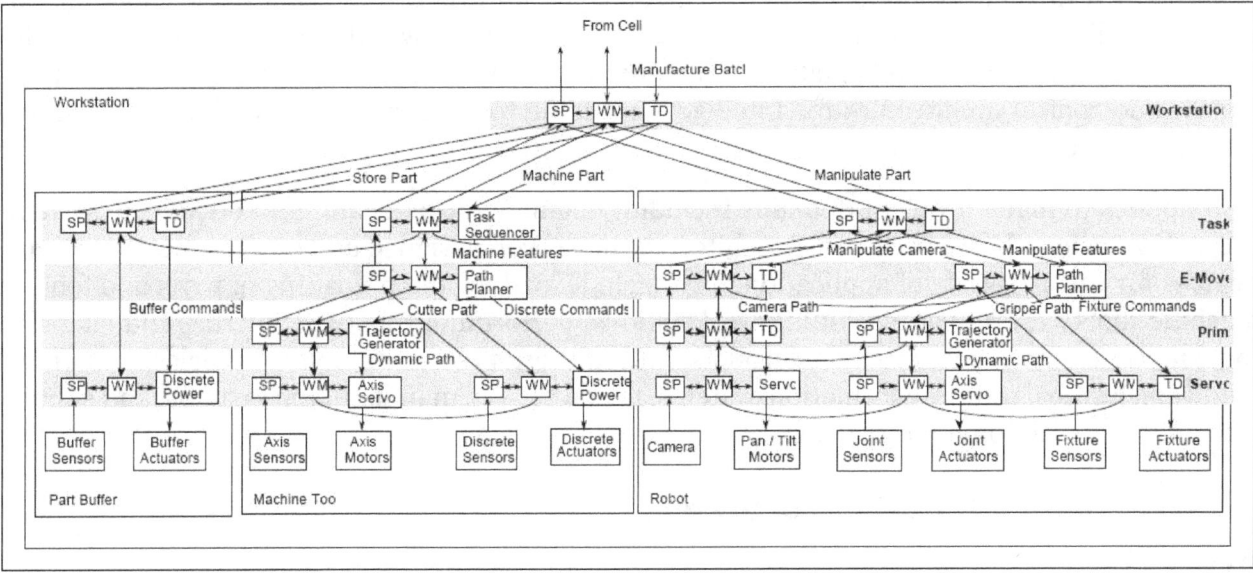

Real-time Control System (RCS) is a reference model architecture, suitable for many software-intensive, real-time control problem domains. RCS is a reference model architecture that defines the types of functions that are required in a real-time intelligent control system, and how these functions are related to each other.

Example of a RCS-3 application of a machining workstation containing a machine tool, part buffer, and robot with vision system. RCS-3 produces a layered graph of processing nodes, each of which contains a task decomposition (TD), world modeling (WM), and sensory processing (SP) module. These modules are richly interconnected to each other by a communications system.

RCS is not a system design, nor is it a specification of how to implement specific systems. RCS prescribes a hierarchical control model based on a set of well-founded engineering principles to organize system complexity. All the control nodes at all levels share a generic node model.

Also RCS provides a comprehensive methodology for designing, engineering, integrating, and testing control systems. Architects iteratively partition system tasks and information into finer, finite subsets that are controllable and efficient. RCS focuses on intelligent control that adapts to uncertain and unstructured operating environments. The key concerns are sensing, perception, knowledge, costs, learning, planning, and execution.

A reference model architecture is a canonical form, not a system design specification. The RCS reference model architecture combines real-time motion planning and control with high level task planning, problem solving, world modeling, recursive state estimation, tactile and visual image processing, and acoustic signature analysis. In fact, the evolution of the RCS concept has been driven by an effort to include the best properties and capabilities of most, if not all, the intelligent control systems currently known in the literature, from subsumption to SOAR, from blackboards to object-oriented programming.

RCS (real-time control system) is developed into an intelligent agent architecture designed to enable any level of intelligent behavior, up to and including human levels of performance. RCS was inspired by a theoretical model of the cerebellum, the portion of the brain responsible for fine motor coordination and control of conscious motions. It was originally designed for sensory-interactive goal-directed control of laboratory manipulators. Over three decades, it has evolved into a real-time control architecture for intelligent machine tools, factory automation systems, and intelligent autonomous vehicles.

RCS applies to many problem domains including manufacturing examples and vehicle systems examples. Systems based on the RCS architecture have been designed and implemented to varying degrees for a wide variety of applications that include loading and unloading of parts and tools in machine tools, controlling machining workstations, performing robotic deburring and chamfering, and controlling space station telerobots, multiple autonomous undersea vehicles, unmanned land vehicles, coal mining automation systems, postal service mail handling systems, and submarine operational automation systems.

Versions of Real-time Control System

RCS-1

In RCS-1, the emphasis was on combining commands with sensory feedback so as to compute the proper response to every combination of goals and states. The application was to control a robot arm with a structured light vision system in visual pursuit tasks. RCS-1 was heavily influenced by biological models such as the Marr-Albus model, and the Cerebellar Model Arithmetic Computer (CMAC) of the cerebellum.

CMAC becomes a state machine when some of its outputs are fed directly back to the input, so RCS-1 was implemented as a set of state-machines arranged in a hierarchy of control levels. At each level, the input command effectively selects a behavior that is driven by feedback in stimulus-response fashion. CMAC thus became the reference model building block of RCS-1, as shown in the figure.

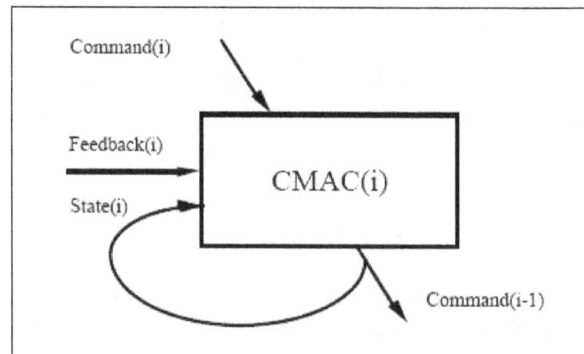

Basics of the RCS-1 control paradigm.

A hierarchy of these building blocks was used to implement a hierarchy of behaviors such as observed by Tinbergen and others. RCS-1 is similar in many respects to Brooks' subsumption architecture, except that RCS selects behaviors before the fact through goals expressed in commands, rather than after the fact through subsumption.

RCS-2

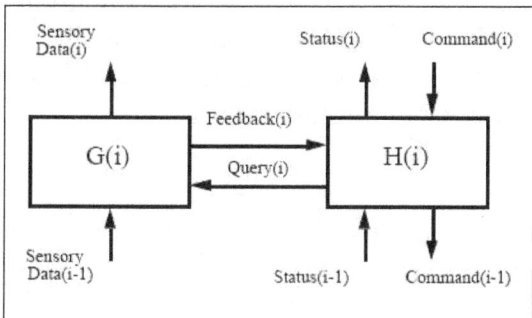

RCS-2 control paradigm.

The next generation, RCS-2, was developed by Barbera, Fitzgerald, Kent, and others for manufacturing control in the NIST Automated Manufacturing Research Facility (AMRF) during the early 1980s. The basic building block of RCS-2 is shown in the figure.

The H function remained a finite state machine state-table executor. The new feature of RCS-2 was the inclusion of the G function consisting of a number of sensory processing algorithms including structured light and blob analysis algorithms. RCS-2 was used to define an eight level hierarchy consisting of Servo, Coordinate Transform, E-Move, Task, Workstation, Cell, Shop, and Facility levels of control.

Only the first six levels were actually built. Two of the AMRF workstations fully implemented five levels of RCS-2. The control system for the Army Field Material Handling Robot (FMR) was also implemented in RCS-2, as was the Army TMAP semi-autonomous land vehicle project.

RCS-3

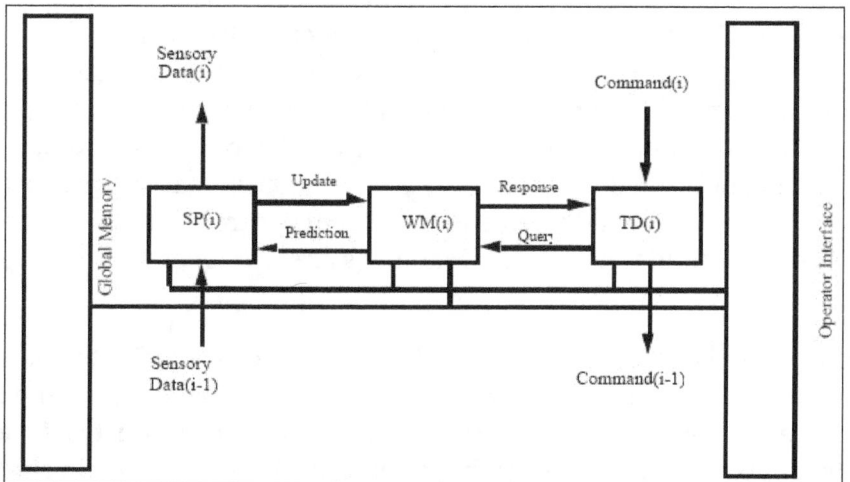

RCS-3 control paradigm.

RCS-3 was designed for the NBS/DARPA Multiple Autonomous Undersea Vehicle (MAUV) project and was adapted for the NASA/NBS Standard Reference Model Telerobot Control System Architecture (NASREM) developed for the space station Flight Telerobotic Servicer The basic building block of RCS-3 is shown in the figure.

The principal new features introduced in RCS-3 are the World Model and the operator interface. The inclusion of the World Model provides the basis for task planning and for model-based sensory processing. This led to refinement of the task decomposition (TD) modules so that each have a job assigner, and planner and executor for each of the subsystems assigned a job. This corresponds roughly to Saridis' three level control hierarchy.

RCS-4

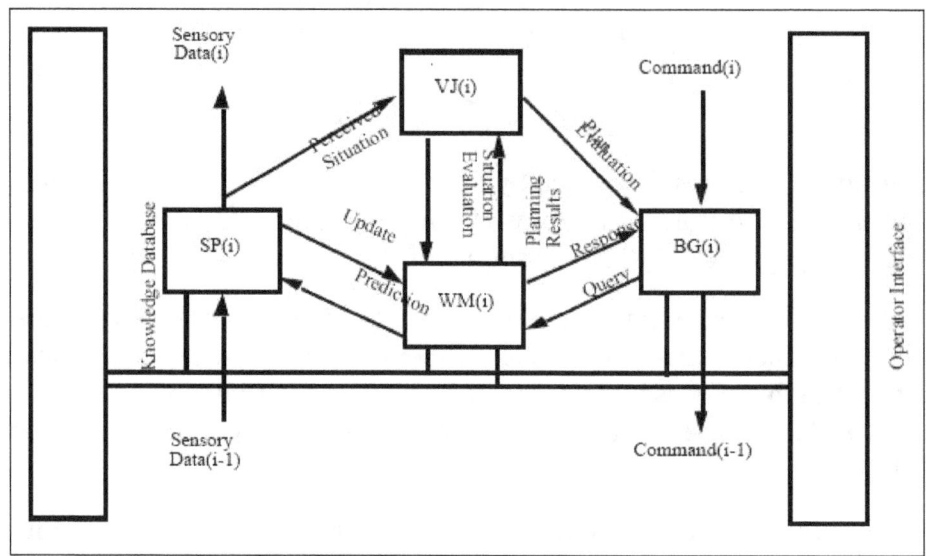

RCS-4 control paradigm.

RCS-4 is developed since the 1990s by the NIST Robot Systems Division. The basic building block is shown in the figure). The principal new feature in RCS-4 is the explicit representation of the Value Judgment (VJ) system. VJ modules provide to the RCS-4 control system the type of functions provided to the biological brain by the limbic system. The VJ modules contain processes that compute cost, benefit, and risk of planned actions, and that place value on objects, materials, territory, situations, events, and outcomes. Value state-variables define what goals are important and what objects or regions should be attended to, attacked, defended, assisted, or otherwise acted upon. Value judgments, or evaluation functions, are an essential part of any form of planning or learning. The application of value judgments to intelligent control systems has been addressed by George Pugh. The structure and function of VJ modules are developed more completely developed in Albus.

RCS-4 also uses the term behavior generation (BG) in place of the RCS-3 term task 5 decomposition (TD). The purpose of this change is to emphasize the degree of autonomous decision making. RCS-4 is designed to address highly autonomous applications in unstructured environments where high bandwidth communications are impossible, such as unmanned vehicles operating on the battlefield, deep undersea, or on distant planets. These applications require autonomous value judgments and sophisticated real-time perceptual capabilities. RCS-3 will continue to be used for less demanding applications, such as manufacturing, construction, or telerobotics for near-space, or shallow undersea operations, where environments are more structured and communication bandwidth to a human interface is less restricted. In these applications, value judgments are often represented implicitly in task planning processes, or in human operator input.

Methodology

In the figure, an example of the RCS methodology for designing a control system for autonomous onroad driving under everyday traffic conditions is summarized in six steps.

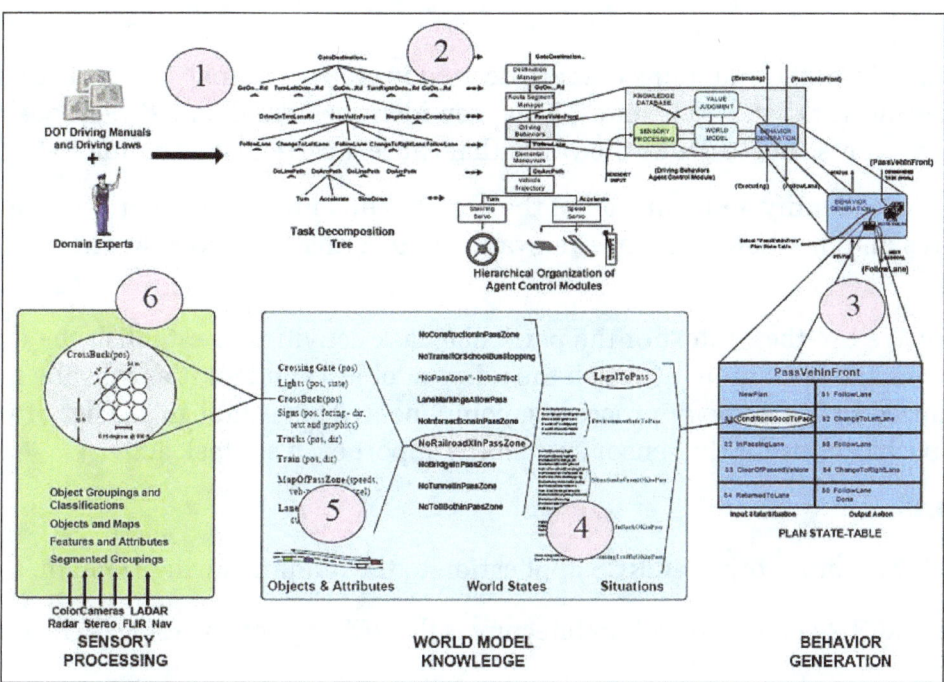

The six steps of the RCS methodology for knowledge acquisition and representation.

- Step 1: Consists of an intensive analysis of domain knowledge from training manuals and subject matter experts. Scenarios are developed and analyzed for each task and subtask. The result of this step is a structuring of procedural knowledge into a task decomposition tree with simpler and simpler tasks at each echelon. At each echelon, a vocabulary of commands (action verbs with goal states, parameters, and constraints) is defined to evoke task behavior at each echelon.

- Step 2: Defines a hierarchical structure of organizational units that will execute the commands defined in step 1. For each unit, its duties and responsibilities in response to each command are specified. This is analogous to establishing a work breakdown structure for a development project, or defining an organizational chart for a business or military operation.

- Step 3: Specifies the processing that is triggered within each unit upon receipt of an input command. For each input command, a state-graph (or statetable or extended finite state automaton) is defined that provides a plan (or procedure for making a plan) for accomplishing the commanded task. The input command selects (or causes to be generated) an appropriate state-table, the execution of which generates a series of output commands to units at the next lower echelon. The library of state-tables contains a set of statesensitive procedural rules that identify all the task branching conditions and specify the corresponding state transition and output command parameters.

The result of step 3 is that each organizational unit has for each input command a state-table of ordered production rules, each suitable for execution by an extended finite state automaton (FSA). The sequence of output subcommands required to accomplish the input command is generated by situations (i.e., branching conditions) that cause the FSA to transition from one output subcommand to the next.

- Step 4: Each of the situations that are defined in step 3 are analyzed to reveal their dependencies on world and task states. This step identifies the detailed relationships between entities, events, and states of the world that cause a particular situation to be true.

- Step 5: We identify and name all of the objects and entities together with their particular features and attributes that are relevant to detecting the above world states and situations.

- Step 6: We use the context of the particular task activities to establish the distances and, therefore, the resolutions at which the relevant objects and entities must be measured and recognized by the sensory processing component. This establishes a set of requirements and specifications for the sensor system to support each subtask activity.

Applications

- The ISAM Framework is an RCS application to the Manufacturing Domain.

- The 4D-RCS Reference Model Architecture is the RCS application to the Vehicle Domain, and

- The NASA/NBS Standard Reference Model for Telerobot Control Systems Architecture (NASREM) is an application to the Space Domain.

ROBUST CONTROL

In control theory, robust control is an approach to controller design that explicitly deals with uncertainty. Robust control methods are designed to function properly provided that uncertain parameters or disturbances are found within some (typically compact) set. Robust methods aim to achieve robust performance and stability in the presence of bounded modelling errors.

The early methods of Bode and others were fairly robust; the state-space methods invented in the 1960s and 1970s were sometimes found to lack robustness, prompting research to improve them. This was the start of the theory of robust control, which took shape in the 1980s and 1990s and is still active today.

In contrast with an adaptive control policy, a robust control policy is static, rather than adapting to measurements of variations, the controller is designed to work assuming that certain variables will be unknown but bounded.

Informally, a controller designed for a particular set of parameters is said to be robust if it also works well under a different set of assumptions. High-gain feedback is a simple example of a robust control method; with sufficiently high gain, the effect of any parameter variations will be negligible. From the closed loop transfer function perspective, high open loop gain leads to substantial disturbance rejection in the face of system parameter uncertainty. Other examples on robust control include sliding mode and terminal sliding mode control.

The major obstacle to achieving high loop gains is the need to maintain system closed loop stability. Loop shaping which allows stable closed loop operation can be a technical challenge.

Robust control systems often incorporate advanced topologies which include multiple feedback loops and feed-forward paths. The control laws may be represented by high order transfer functions required to simultaneously accomplish desired disturbance rejection performance with robust closed loop operation.

High-gain feedback is the principle that allows simplified models of operational amplifiers and emitter-degenerated bipolar transistors to be used in a variety of different settings. This idea was already well understood by Bode and Black in 1927.

The Modern Theory of Robust Control

The theory of robust control began in the late 1970s and early 1980s and soon developed a number of techniques for dealing with bounded system uncertainty.

Probably the most important example of a robust control technique is H-infinity loop-shaping, which was developed by Duncan McFarlane and Keith Glover of Cambridge University; this method minimizes the sensitivity of a system over its frequency spectrum, and this guarantees that the system will not greatly deviate from expected trajectories when disturbances enter the system.

An emerging area of robust control from application point of view is sliding mode control (SMC), which is a variation of variable structure control (VSC). The robustness properties of SMC with respect to matched uncertainty as well as the simplicity in design attracted a variety of applications.

While robust control has been traditionally dealt with along deterministic approaches, in the last two decades this approach has been criticized on the basis that it is too rigid to describe real uncertainty, while it often also leads to over conservative solutions. Probabilistic robust control has been introduced as an alternative, see e.g. that interprets robust control within the so-called scenario optimization theory.

Another example is loop transfer recovery (LQG/LTR), which was developed to overcome the robustness problems of linear-quadratic-Gaussian control (LQG) control.

Other robust techniques includes quantitative feedback theory (QFT), passivity based control, Lyapunov based control, etc.

When system behavior varies considerably in normal operation, multiple control laws may have to be devised. Each distinct control law addresses a specific system behavior mode. An example is a computer hard disk drive. Separate robust control system modes are designed in order to address the rapid magnetic head traversal operation, known as the seek, a transitional settle operation as the magnetic head approaches its destination, and a track following mode during which the disk drive performs its data access operation.

One of the challenges is to design a control system that addresses these diverse system operating modes and enables smooth transition from one mode to the next as quickly as possible.

Such state machine driven composite control system is an extension of the gain scheduling idea where the entire control strategy changes based upon changes in system behavior.

STOCHASTIC CONTROL

Stochastic control or stochastic optimal control is a sub field of control theory that deals with the existence of uncertainty either in observations or in the noise that drives the evolution of the system. The system designer assumes, in a Bayesian probability-driven fashion, that random noise with known probability distribution affects the evolution and observation of the state variables. Stochastic control aims to design the time path of the controlled variables that performs the desired control task with minimum cost, somehow defined, despite the presence of this noise. The context may be either discrete time or continuous time.

Certainty Equivalence

An extremely well-studied formulation in stochastic control is that of linear quadratic Gaussian control. Here the model is linear, the objective function is the expected value of a quadratic form, and the disturbances are purely additive. A basic result for discrete-time centralized systems with only additive uncertainty is the certainty equivalence property: that the optimal control solution in this case is the same as would be obtained in the absence of the additive disturbances. This property is applicable to all centralized systems with linear equations of evolution, quadratic cost function, and noise entering the model only additively; the quadratic assumption allows for the optimal control laws, which follow the certainty-equivalence property, to be linear functions of the observations of the controllers.

Any deviation from the above assumptions—a nonlinear state equation, a non-quadratic objective function, noise in the multiplicative parameters of the model, or decentralization of control—causes the certainty equivalence property not to hold. For example, its failure to hold for decentralized control was demonstrated in Witsenhausen's counterexample.

Discrete Time

In a discrete-time context, the decision-maker observes the state variable, possibly with observational noise, in each time period. The objective may be to optimize the sum of expected values of a nonlinear (possibly quadratic) objective function over all the time periods from the present to the final period of concern, or to optimize the value of the objective function as of the final period only. At each time period new observations are made, and the control variables are to be adjusted optimally. Finding the optimal solution for the present time may involve iterating a matrix Riccati equation backwards in time from the last period to the present period.

In the discrete-time case with uncertainty about the parameter values in the transition matrix (giving the effect of current values of the state variables on their own evolution) and the control response matrix of the state equation, but still with a linear state equation and quadratic objective function, a Riccati equation can still be obtained for iterating backward to each period's solution even though certainty equivalence does not apply. The discrete-time case of a non-quadratic loss function but only additive disturbances can also be handled, albeit with more complications.

Example:

A typical specification of the discrete-time stochastic linear quadratic control problem is to minimize,

$$\mathrm{E}_1 \sum_{t=1}^{S} \left[y_t^T Q y_t + u_t^T R u_t \right]$$

where E_1 is the expected value operator conditional on y_0, superscript T indicates a matrix transpose, and S is the time horizon, subject to the state equation,

$$y_t = A_t y_{t-1} + B_t u_t,$$

where y is an $n \times 1$ vector of observable state variables, u is a $k \times 1$ vector of control variables, A_t is the time t realization of the stochastic $n \times n$ state transition matrix, B_t is the time t realization of the stochastic $n \times k$ matrix of control multipliers, and Q $(n \times n)$ and R $(k \times k)$ are known symmetric positive definite cost matrices. We assume that each element of A and B is jointly independently and identically distributed through time, so the expected value operations need not be time-conditional.

Induction backwards in time can be used to obtain the optimal control solution at each time,

$$u_t^* = -[\mathrm{E}(B^T X_t B + R)]^{-1} \mathrm{E}(B^T X_t A) y_{t-1},$$

with the symmetric positive definite cost-to-go matrix X evolving backwards in time from $X_S = Q$ according to,

$$X_{t-1} = Q + \mathrm{E}[A^T X_t A] - \mathrm{E}[A^T X_t B][\mathrm{E}(B^T X_t B + R)]^{-1} \mathrm{E}(B^T X_t A),$$

which is known as the discrete-time dynamic Riccati equation of this problem. The only information needed regarding the unknown parameters in the A and B matrices is the expected value and variance of each element of each matrix and the covariances among elements of the same matrix and among elements across matrices.

The optimal control solution is unaffected if zero-mean, i.i.d. additive shocks also appear in the state equation, so long as they are uncorrelated with the parameters in the A and B matrices. But if they are so correlated, then the optimal control solution for each period contains an additional additive constant vector. If an additive constant vector appears in the state equation, then again the optimal control solution for each period contains an additional additive constant vector.

The steady-state characterization of X (if it exists), relevant for the infinite-horizon problem in which S goes to infinity, can be found by iterating the dynamic equation for X repeatedly until it converges; then X is characterized by removing the time subscripts from its dynamic equation.

Continuous Time

If the model is in continuous time, the controller knows the state of the system at each instant of time. The objective is to maximize either an integral of, for example, a concave function of a state variable over a horizon from time zero (the present) to a terminal time T, or a concave function of a state variable at some future date T. As time evolves, new observations are continuously made and the control variables are continuously adjusted in optimal fashion.

OPTIMAL CONTROL

Optimal control is the process of determining control and state trajectories for a dynamic system over a period of time to minimise a performance index.

Formulation of optimal control problems

There are various types of optimal control problems, depending on the performance index, the type of time domain (continuous, discrete), the presence of different types of constraints, and what variables are free to be chosen. The formulation of an optimal control problem requires the following:

- A mathematical model of the system to be controlled,

- A specification of the performance index,

- A specification of all boundary conditions on states, and constraints to be satisfied by states and controls,

- A statement of what variables are free.

Continuous Time Optimal Control Using the Variational Approach

General Case with Fixed Final Time and no Terminal or Path Constraints

If there are no path constraints on the states or the control variables, and if the initial and final times are fixed, a fairly general continuous time optimal control problem can be defined as follows:

Problem 1: Find the control vector trajectory $u:[t_0, t_f] \subset \mathbb{R} \mapsto \mathbb{R}^{n_u}$ to minimize the performance index:

$$J = \varphi\left(x\left(t_f\right)\right) + \int_{t_0}^{t_f} L\left(x(t), u(t), t\right) dt$$

subject to:

$$\dot{x}(t) = f\left(x(t), u(t), t\right), \ x\left(t_0\right) = x_0$$

Where $[t_0, t_f]$ is the time interval of interest, $x:[t_0, t_f] \mapsto \mathbb{R}^{n_x}$ is the state vector, $\varphi:\mathbb{R}^{n_x} \times \mathbb{R} \mapsto \mathbb{R}$ is a terminal cost function, $L:\mathbb{R}^{n_x} \times \mathbb{R}^{n_u} \times \mathbb{R} \mapsto \mathbb{R}$ is an intermediate cost function, and

$f : \mathbb{R}^{n_x} \times \mathbb{R}^{n_u} \times \mathbb{R} \mapsto \mathbb{R}^{n_z}$ is a vector field. Note that equation $\dot{x}(t) = f\big(x(t), u(t), t\big)$, $x(t_0) = x_0$ represents the dynamics of the system and its initial state condition. Problem 1 as defined above is known as the Bolza problem. If L (x, u, t) = 0 , then the problem is known as the Mayer problem, if $\varphi\big(x(t_f)\big) = 0$, it is known as the Lagrange problem. Note that the performance index $J = J$ (u) is a functional, this is a rule of correspondence that assigns a real value to each function u in a class. Calculus of variations is concerned with the optimisation of functionals, and it is the tool that is used in this section to derive necessary optimality conditions for the minimisation of J (u).

Adjoin the constraints to the performance index with a time-varying Lagrange multiplier vector function $\lambda : [t_0, t_f] \mapsto \mathbb{R}^{n_x}$ (also known as the co-state), to define an augmented performance index $\bar{J}$:

$$\bar{J} = \varphi\big(x(t_f)\big) + \int_{t_0}^{t_f} \Big\{ L\big(x, u, t\big) + \lambda^T(t) \big[f(x, u, t) - \dot{x} \big] \Big\} dt$$

Define the Hamiltonian function H as follows:

$$H\big(x(t), u(t), \lambda(t), t\big) = L\big(x(t), u(t), t\big) + \lambda(t)^T f\big(x(t), u(t), t\big),$$

such that $\bar{J}$ can be written as:

$$\bar{J} = \varphi\big(x(t_f)\big) + \int_{t_0}^{t_f} \Big\{ H\big(x(t), u(t), \lambda(t), t\big) - \lambda^T(t)\dot{x} \Big\} dt$$

Assume that t_0 and t_f are fixed. Now consider an infinitesimal variation in u(t) , that is denoted as δu(t). Such a variation will produce variations in the state history δx(t) , and a variation in the performance index $\delta\bar{J}$:

$$\delta\bar{J} = \left[\left(\frac{\partial\varphi}{\partial x} - \lambda^T \right) \delta x \right]_{t=t_f} + \left[\lambda^T \delta x \right]_{t=t_0} + \int_{t_0}^{t_f} \left\{ \left(\frac{\partial H}{\partial x} + \dot{\lambda}^T \right) \delta x + \left(\frac{\partial H}{\partial u} \right) \delta u \right\} dt$$

Since the Lagrange multipliers are arbitrary, they can be selected to make the coefficients of δx(t) and δx(t_f) equal to zero, as follows:

$$\dot{\lambda}^T = -\frac{\partial H}{\partial x},$$

$$\lambda(t_f)^T = \frac{\partial\varphi}{\partial x}\bigg|_{t=t_f}.$$

This choice of $\lambda(t)$ results in the following expression for $\bar{J}$, assuming that the initial state is fixed, so that δx(t_0)=0 :

$$\delta\bar{J} = \int_{t_0}^{t_f} \left\{ \left(\frac{\partial H}{\partial u} \right) \delta u \right\} dt$$

For a minimum, it is necessary that $\delta \bar{J} = 0$. This gives the stationarity condition:

$$\frac{\partial H^T}{\partial u} = 0.$$

Equations $\dot{x}(t) = f(x(t), u(t), t)$, $x(t_0) = x_0$, $\lambda^T = -\dfrac{\partial H}{\partial x}$, $\lambda(t_f)^T = \dfrac{\partial \varphi}{\partial x}\bigg|_{t=t_f}$, and $\dfrac{\partial H^T}{\partial u} = 0$ are the

first-order necessary conditions for a minimum of J. Equation $\lambda^T = -\dfrac{\partial H}{\partial x}$, is known as the co-state

(or adjoint) equation. Equation $\lambda(t_f)^T = \dfrac{\partial \varphi}{\partial x}\bigg|_{t=t_f}$ and the initial state condition represent the

boundary (or transversality) conditions. These necessary optimality conditions, which define a two point boundary value problem, are very useful as they allow to find analytical solutions to special types of optimal control problems, and to define numerical algorithms to search for solutions in general cases. Moreover, they are useful to check the extremality of solutions found by computational methods. Sufficient conditions for general nonlinear problems have also been established. Distinctions are made between sufficient conditions for weak local, strong local, and strong global minima. Sufficient conditions are useful to check if an extremal solution satisfying the necessary optimality conditions actually yields a minimum, and the type of minimum that is achieved.

The Linear Quadratic Regulator

A special case of optimal control problem which is of particular importance arises when the objective function is a quadratic function of x and u, and the dynamic equations are linear. The resulting feedback law in this case is known as the linear quadratic regulator (LQR). The performance index is given by:

$$J = \frac{1}{2} x(t_f)^T S_f x(t_f) + \frac{1}{2} \int_{t_0}^{t_f} \left(x(t)^T Q x(t) + u(t)^T R u(t) \right) dt$$

where S_f and Q are positive semidefinite matrices, and R is a positive definite matrix, while the system dynamics obey:

$$\dot{x}(t) = Ax(t) Bu(t), \; x(t_0) = x_0$$

where A is the system matrix and B is the input matrix.

In this case, using the optimality conditions given above, it is possible to find that the optimal control law can be expressed as a linear state feedback:

$$u(t) = -K(t)x(t)$$

where the state feedback gain is given by:

$$K(t) = R^{-1} B^T S(t),$$

and S(t) is the solution to the differential Ricatti equation

$$-\dot{S} = A^T S + SA - SBR^{-1} B^T S + Q, \; S(t_f) = S_f$$

In the particular case where tf→∞ , and provided the pair (A, B) is stabilizable, the Ricatti differential equation converges to a limiting solution S, and it is possible to express the optimal control law as a state feedback as in $u(t) = -K(t)x(t)$ but with constant gain K. which is given by

$$K = R^{-1} B^T S$$

where S is the positive definite solution to the algebraic Ricatti equation:

$$A^T S + SA - SBR^{-1} B^T S + Q = 0$$

Moreover, if the pair (A,C) is observable, where $C^T C = Q$, then the closed loop system

$$\dot{x} = (A - BK)x$$

is asymptotically stable. This is an important result, as the linear quadratic regulator provides a way of stabilizing any linear system that is stabilizable. It is worth pointing out that there are well established methods and software for solving the algebraic Ricatti equation $A^T S + SA - SBR^{-1} B^T S + Q = 0$. This facilitates the design of linear quadratic regulators. A useful extension of the linear quadratic regulator ideas involves modifying the performance index

$$J = \frac{1}{2}x(t_f)^T S_f x(t_f) + \frac{1}{2}\int_{t_0}^{t_f} \left(x(t)^T Qx(t) + u(t)^T Ru(t) \right) dt$$ to allow for a reference signal that the

output of the system should track. Moreover, an extension of the LQR concept to systems with gaussian additive noise, which is known as the linear quadratic gaussian (LQG) controller, has been widely applied. The LQG controller involves coupling the linear quadratic regulator with the Kalman filter using the separation principle.

Case with Terminal Constraints

In case problem 1 is also subject to a set of terminal constraints of the form:

$$\psi\left(x(t_f), t_f\right) = 0$$

Where $\psi : \mathbb{R}^{n_x} \times \mathbb{R} \mapsto \mathbb{R}^{n_\psi}$ is a vector function, variational analysis shows that the necessary conditions for a minimum of J are $\frac{\partial H^T}{\partial u} = 0$, $\dot{\lambda}^T = -\frac{\partial H}{\partial x}$, $\dot{x}(t) = f\left(x(t), u(t), t\right)$, $x(t_0) = x_0$ and the following terminal condition:

$$\left(\frac{\partial \varphi^T}{\partial x} + \frac{\partial \psi^T}{\partial x} v - \lambda \right)^T \Bigg|_{t_f} \delta x(t_f) + \left(\frac{\partial \varphi}{\partial t} + \frac{\partial \psi^T}{\partial t} v + H \right) \Bigg|_{t_f} \delta t_f = 0$$

where $v \in \mathbb{R}^{n_\psi}$ is the Lagrange multiplier associated with the terminal constraint, δt_f is the variation of the final time, and $\delta x(t_f)$ is the variation of the final state. Note that if the final time is fixed, then $\delta t_{f=0}$ and the second term vanishes. Also, if the terminal constraint is such that element j of x is fixed at the final time, then element j of $\delta x(t_f)$ vanishes.

Case with Input Constraints - The Minimum Principle

Realistic optimal control problems often have inequality constraints associated with the input variables, so that the input variable u is restricted to be within an admissible compact region Ω, such that:

$$u(t) \in \Omega .$$

It was shown by Pontryagin and co-workers (Pontryagin, 1987) that in this case, the necessary conditions $\dot{x}(t) = f\left(x(t), u(t), t\right)$, $x(t_0) = x_0$, $\dot{\lambda}^T = -\dfrac{\partial H}{\partial x}$, and $\lambda(t_f)^T = \dfrac{\partial \varphi}{\partial x}\Big|_{t=t_f}$ still hold, but the stationarity condition $\dfrac{\partial H^T}{\partial u} = 0$, has to be replaced by:

$$H\left(x^*(t), u^*(t), \lambda^*(t), t\right) \leq H\left(x^*(t), u(t), \lambda^*(t), t\right)$$

for all admissible u, where * denotes optimal variables. This condition is known as Pontryagin's *minimum principle*. According to this principle, the Hamiltonian must be minimised over all admissible u for optimal values of the state and costate variables.

Minimum Time Problems

One special class of optimal control problem involves finding the optimal input $u(t)$ to reach a terminal constraint in minimum time. This kind of problem is defined as follows.

Problem 2: Find t_f and $u(t)(t \in [t_0, t_f])$ to minimise:

$$J = \int_{t_0}^{t_f} 1 dt = t_f - t_0$$

subject to:

$$\dot{x}(t) = f(x(t), u(t), t), \ x(0) = x_0$$

$$\psi\left(x(t_f), t_f\right) = 0$$

$$u(t) \in \Omega.$$

Problems with Path Constraints

Sometimes it is necessary to restrict state and control trajectories such that a set of constraints is satisfied within the interval of interest $[t_0, t_f]$:

$$c(x(t), u(t), t) \leq 0$$

where $c: \mathbb{R}^{n_x} \times \mathbb{R}^{n_u} \times \left[t_0, t_f\right] \mapsto \mathbb{R}^{n_c}$. Moreover, in some problems it may be required that the state satisfies equality constraints at some intermediate point in time $t_1, t_0 \leq t_1 \leq t_f$. These are known as interior point constraints and can be expressed as follows:

$$q(x(t_1), t_1) = 0$$

where $q : \mathbb{R}^{n_z} \times \mathbb{R} \mapsto \mathbb{R}^{n_q}$.

Singular Arcs

In some optimal control problems, extremal arcs satisfying $\dfrac{\partial H^T}{\partial u} = 0$ occur where the matrix $\partial^2 H / \partial u^2$ is singular. These are called singular arcs. Additional tests are required to verify if a singular arc is optimizing. A particular case of practical relevance occurs when the Hamiltonian function is linear in at least one of the control variables. In such cases, the control is not determined in terms of the state and co-state by the stationarity condition $\dfrac{\partial H^T}{\partial u} = 0$. Instead, the control is determined by the condition that the time derivatives of $\partial H / \partial u$ must be zero along the singular arc. In the case of a single control u, once the control is obtained by setting the time derivative of $\partial H / \partial u$ to zero, then additional necessary conditions known as the generalized Legendre-Clebsch conditions must be checked:

$$(-1)^k \frac{\partial}{\partial u} \left[\frac{d^{(2k)}}{dt^{2k}} \frac{\partial H}{\partial u} \right] \geq 0, k = 0, 1, 2, \dots$$

The presence of singular arcs may cause difficulties to computational optimal control methods to find accurate solutions if the appropriate conditions are not enforced a priori.

Computational Optimal Control

The solutions to many optimal control problems cannot be found by analytical means. Over the years, many numerical procedures have been developed to solve general optimal control problems. With direct methods, optimal control problems are discretised and converted into nonlinear programming problems of the form:

Problem 3: Find a decision vector $y \in \mathbb{R}^{n_y}$ to minimise $F(y)$ subject to $g(y) \leq 0$, $h(y) = 0$, and simple bounds $y_l \leq y \leq y_u$, where $F : \mathbb{R}^{n_y} \mapsto \mathbb{R}$ is a differentiable scalar function, $g : \mathbb{R}^{n_y} \mapsto \mathbb{R}^{n_g}$ and $h : \mathbb{R}^{n_y} \mapsto \mathbb{R}^{n_h}$ are differentiable vector functions. Some methods involve the discretization of the differential equations using, for example, Euler, Trapezoidal, or Runge-Kutta methods, by defining a grid of N points covering the time interval $[t_0, t_f]$, $t_0 = t_1 < t_2 \dots < t_N = t_f$. In this way, the differential equations become equality constraints of the nonlinear programming problem. The decision vector y contains the control and state variables at the grid points. Other direct methods involve a decision vector y which contains only the control variables at the grid points, with the differential equations solved by integration and their gradients found by integrating the co-state equations, or by finite differences. Other direct methods involve the approximation of the control and states using basis functions, such as splines or Lagrange polynomials. There are well established numerical techniques for solving nonlinear programming problems with constraints, such as sequential quadratic programming. Direct methods using nonlinear programming are known to deal in an efficient manner with problems involving path constraints.

Indirect methods involve iterating on the necessary optimality conditions to seek their satisfaction. This usually involves attempting to solve nonlinear two-point boundary value problems, through the forward integration of the plant equations and the backward integration of the co-state equations. Examples of indirect methods include the gradient method and the multiple shooting method.

Dynamic Programming

Dynamic programming is an alternative to the variational approach to optimal control. It was proposed by Bellman in the 1950s, and is an extension of Hamilton-Jacobi theory. Bellman's *principle of optimality* is stated as follows: "An optimal policy has the property that regardless of what the previous decisions have been, the remaining decisions must be optimal with regard to the state resulting from those previous decisions". This principle serves to limit the number of potentially optimal control strategies that must be investigated. It also shows that the optimal strategy must be determined by working backward from the final time.

Consider Problem 1 with the addition of a terminal state constraint $\psi\left(x\left(t_f\right), t_f\right) = 0$. Using Bellman's principle of optimality, it is possible to derive the Hamilton-Jacobi-Bellman (HJB) equation:

$$-\frac{\partial J^*}{\partial t} = \min_u \left(L + \frac{\partial J^*}{\partial x} f \right)$$

where J^* is the optimal performance index. In some cases, the HJB equation can be used to find analytical solutions to optimal control problems.

Dynamic programming includes formulations for discrete time systems as well as combinatorial systems, which are discrete systems with quantized states and controls. Discrete dynamic programming, however, suffers from the 'curse of dimensionality', which causes the computations and memory requirements to grow dramatically with the problem size.

Discrete-time Optimal Control

Most of the problems defined above have discrete-time counterparts. These formulations are useful when the dynamics are discrete (for example, a multistage system), or when dealing with computer controlled systems. In discrete-time, the dynamics can be expressed as a difference equation:

$x(k+1) = f(x(k), u(k), k), x(N_o) = x_o$

where k is an integer index, $x(k)$ is the state vector, $u(k)$ is the control vector, and f is a vector function. The objective is to find a control sequence $\{u(k)\}, k = N_{0,...,} N_f - 1$, to minimise a performance index of the form:

$$J = \varphi\left(x\left(N_f\right)\right) + \sum_{k=N_0}^{N_f - 1} L(x(k), u(k), k)$$

Examples:

Minimum Energy Control of a Double Integrator with Terminal Constraint

Consider the following optimal control problem.

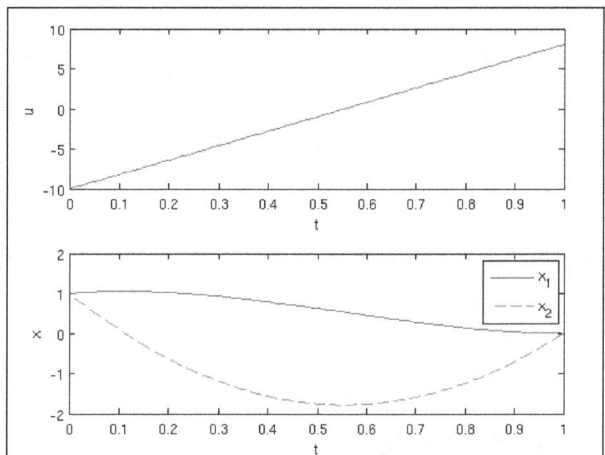

Optimal control and state histories for the double integrator example.

$$\min_{u(t)} J = \int_0^1 u(t)^2 \, dt$$

subject to,

$$\dot{x}_1(t) = x_2(t), \dot{x}_2(t) = u(t),$$

$$x_1(0) = 1, \ x_2(0) = 1, \ x_1(1) = 0, \ x_2(1) = 0$$

The Hamiltonian function $H(x(t), u(t), \lambda(t), t) = L(x(t), u(t), t) + \lambda(t)^T f(x(t), u(t), t)$ is given by:

$$H = \frac{1}{2}u^2 + \lambda_1 x_2 + \lambda_2 u$$

The stationarity condition $\dfrac{\partial H^T}{\partial u} = 0$ yields:

$$u + \lambda_2 = 0 \Rightarrow u = -\lambda_2$$

The co-state equation $\dot{\lambda}^T = -\dfrac{\partial H}{\partial x}$, gives:

$$\dot{\lambda}_1 = 0, \dot{\lambda}_2 = -\lambda_1,$$

so that,

$$\lambda_1(t) = a, \lambda_2(t) = -at + b,$$

where a and b are constants to be found. Replacing $\lambda_1(t) = a$, $\lambda_2(t) = -at + b$, in $u + \lambda_2 = 0 \Rightarrow u = -\lambda_2$ gives:

$$u(t) = at - b.$$

In this case, the terminal constraint function is $\psi(x(1)) = [x_1(1), x_2(1)]^T = [0, 0]^T$, so that the final value of the state vector is fixed, which implies that $\delta x(t_f) = 0$. Noting that $\delta t_f = 0$ since the final time is fixed, then the terminal condition $\left(\dfrac{\partial \varphi^T}{\partial \mathrm{x}} + \dfrac{\partial \psi^T}{\partial \mathrm{x}} v - \lambda\right)^T\bigg|_{t_f} \delta \mathrm{x}(t_f) + \left(\dfrac{\partial \varphi}{\partial t} + \dfrac{\partial \psi^T}{\partial t} v + H\right)\bigg|_{t_f} \delta t_f = 0$

is satisfied. Replacing $u(t) = at - b$ into the state equation $\dot{x}_1(t) = x_2(t), \dot{x}_2(t) = u(t)$, and integrating both states gives:

$$x_1(t) = \frac{1}{6}at^3 - \frac{1}{2}bt^2 + ct + d, x_2(t) = \frac{1}{2}at^2 - bt + c.$$

Evaluating $x_1(t) = \frac{1}{6}at^3 - \frac{1}{2}bt^2 + ct + d, x_2(t) = \frac{1}{2}at^2 - bt + c$ at $t=0$ and using the initial conditions

gives the values $c=1$ and $d=1$. Evaluating $x_1(t) = \frac{1}{6}at^3 - \frac{1}{2}bt^2 + ct + d, x_2(t) = \frac{1}{2}at^2 - bt + c$ at the

terminal time $t=1$ gives two simultaneous equations:

$$\frac{1}{6}a - \frac{1}{2}b + 2 = 0, \frac{1}{2}a - b + 1 = 0.$$

This yields $a=18$, and $b=10$. Therefore, the optimal control is given by:

$$u = 18t - 10.$$

Computational Optimal Control: B-727 Maximum Altitude Climbing Turn Manoeuvre

This example is solved using a gradient method in. Here, a path constraint is considered and the solution is sought by using a direct method and nonlinear programming. It is desired to find the optimal control histories to maximise the altitude of a B-727 aircraft in a given time t_f, with terminal constraints that the aircraft path be turned 60 degrees and the velocity be slightly above the stall velocity. Such a flight path may be of interest to reduce engine noise over populated areas located ahead of an airport runway. This manoeuvre can be formulated as an optimal control problem, as follows.

$$\min_{u(t), a(t)} J = -h(t_f)$$

subject to:

$$\dot{V} = T(V)\cos(\alpha + \varepsilon) - C_D(\alpha)V^2 - \sin\gamma,$$

$$\dot{\gamma} = (1/V)\left[T(V)\sin(\alpha + \varepsilon) + C_L(\alpha)V^2\right]\cos\sigma - (1/V)\cos\gamma,$$

$$\dot{\psi} = (1/(V\cos\gamma))\left[T(V)\sin(\alpha + \varepsilon) + C_L(\alpha)V^2\right]\sin\sigma,$$

$$\dot{h} = V\sin\gamma,$$

$$\dot{x} = V\cos\gamma\cos\psi,$$

$$\dot{y} = V\cos\gamma\sin\psi.$$

with initial conditions given by:

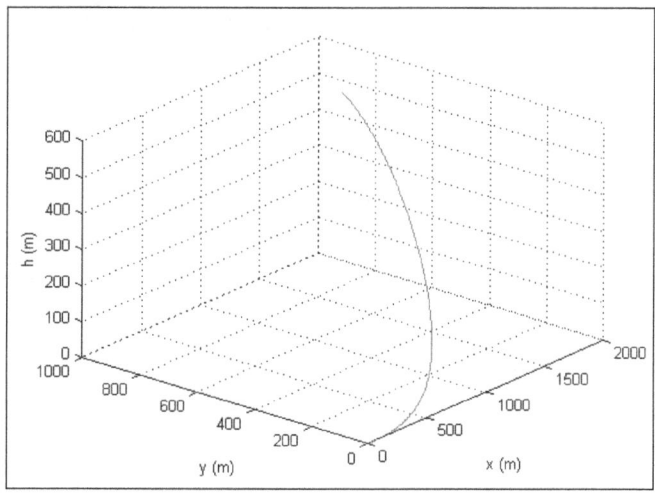

3D plot of optimal B-727 aircraft trajectory.

$V(0) = 1.0$

$\gamma(0) = \psi(0) = h(0) = x(0) = y(0) = 0$

tbe terminal constraints:

$$V\left(t_f\right) = 0.60, \; \psi\left(t_f\right) = \left.\frac{\pi}{3}\right|$$

and the path constraint:

$h(t) \geq 0, \; t \in [0, t_f]$

where h is the altitude, x is the horizontal distance in the initial direction, y is the horizontal distance perpendicular to the initial direction, V is the aircraft velocity, γ is the climb angle, ψ is the heading angle, and $t_f = 2.4$ units. The distance and time units in the above equations are normalised. To obtain meters and seconds, the corresponding variables need to be multiplied by 10.0542, and 992.0288, respectively. There are two controls: the angle of attack α and the bank angle σ. The functions $T(V)$, $C_D(\alpha)$ and $C_L(\alpha)$ are given by:

$$T\left(V\right) = 0.2476 - 0.04312V + 0.008392V^2$$

$$C_D\left(\alpha\right) = 0.07351 - 0.08617\alpha + 1.996\alpha^2$$

$$C_L\left(\alpha\right) = \begin{cases} 0.1667 + 6.231\alpha, & \text{if } \alpha \leq 12\pi/180 \\ 0.1667 + 6.231\alpha + 21.65\left(a - 12\pi/180\right)^2 & \text{if } \alpha \leq 12\pi/180 \end{cases}$$

The solution shown in figure was obtained by using sequential quadratic programming, where the decision vector consisted of the control values at the grid points. The differential equations were integrated using 5th order Runge-Kutta steps with size $\Delta t = 0.01$ units, and the gradients required by the nonlinear programming code were found by finite differences.

References

- What-is-adaptive-control: automationforum.co, Retrieved 19 February, 2019

- Vichuzhanin, Vladimir (12 April 2012). "Realization of a fuzzy controller with fuzzy dynamic correction". Central European Journal of Engineering. 2 (3): 392–398. doi:10.2478/s13531-012-0003-7

- Optimal-control: scholarpedia.org, Retrieved 23 June, 2019

- M. Sami Fadali, Antonio Visioli, (2009) "Digital Control Engineering", Academic Press, ISBN 978-0-12-374498-2

- Khalil, H. K. (2002). Nonlinear Systems (3rd ed.). Upper Saddle River: Prentice Hall. ISBN 978-0-13-067389-3

Control Systems

Control Systems can be defined as the systems which manage and regulate the behavior of other systems. They can be categorized into open-loop systems, closed-loop systems, feedback systems, negative feedback systems, etc. This chapter has been carefully written to provide an easy understanding of these varied facets of control systems as well as their advantages and disadvantages.

A control system is a system, which provides the desired response by controlling the output. The following figure shows the simple block diagram of a control system.

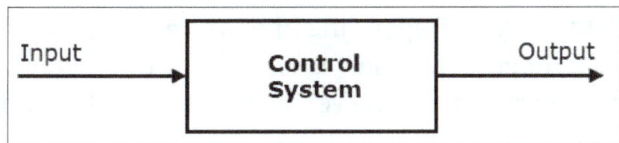

Here, the control system is represented by a single block. Since, the output is controlled by varying input, the control system got this name. We will vary this input with some mechanism.

Examples – Traffic lights control system, washing machine

Traffic lights control system is an example of control system. Here, a sequence of input signal is applied to this control system and the output is one of the three lights that will be on for some duration of time. During this time, the other two lights will be off. Based on the traffic study at a particular junction, the on and off times of the lights can be determined. Accordingly, the input signal controls the output. So, the traffic lights control system operates on time basis.

Classification of Control Systems

Based on some parameters, we can classify the control systems into the following ways.

Continuous time and Discrete-time Control Systems:

- Control Systems can be classified as continuous time control systems and discrete time control systems based on the type of the signal used.

- In continuous time control systems, all the signals are continuous in time. But, in discrete time control systems, there exists one or more discrete time signals.

SISO and MIMO Control Systems

- Control Systems can be classified as SISO control systems and MIMO control systems based on the number of inputs and outputs present.

- SISO (Single Input and Single Output) control systems have one input and one output. Whereas, MIMO (Multiple Inputs and Multiple Outputs) control systems have more than one input and more than one output.

OPEN-LOOP SYSTEM

The function of any electronic system is to automatically regulate the output and keep it within the systems desired input value or "set point". If the systems input changes for whatever reason, the output of the system must respond accordingly and change itself to reflect the new input value.

Likewise, if something happens to disturb the systems output without any change to the input value, the output must respond by returning back to its previous set value. In the past, electrical control systems were basically manual or what is called an Open-loop System with very few automatic control or feedback features built in to regulate the process variable so as to maintain the desired output level or value.

For example, an electric clothes dryer. Depending upon the amount of clothes or how wet they are, a user or operator would set a timer (controller) to say 30 minutes and at the end of the 30 minutes the drier will automatically stop and turn-off even if the clothes where still wet or damp.

In this case, the control action is the manual operator assessing the wetness of the clothes and setting the process (the drier) accordingly.

So in this example, the clothes dryer would be an open-loop system as it does not monitor or measure the condition of the output signal, which is the dryness of the clothes. Then the accuracy of the drying process, or success of drying the clothes will depend on the experience of the user (operator).

However, the user may adjust or fine tune the drying process of the system at any time by increasing or decreasing the timing controllers drying time, if they think that the original drying process will not be met. For example, increasing the timing controller to 40 minutes to extend the drying process. Consider the following open-loop block diagram.

Open-loop Drying System

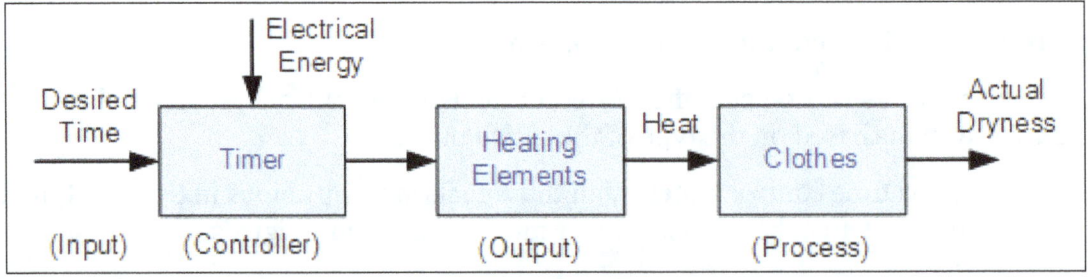

Then an Open-loop system, also referred to as non-feedback system, is a type of continuous control system in which the output has no influence or effect on the control action of the input signal.

In other words, in an open-loop control system the output is neither measured nor "fed back" for comparison with the input. Therefore, an open-loop system is expected to faithfully follow its input command or set point regardless of the final result.

Also, an open-loop system has no knowledge of the output condition so cannot self-correct any errors it could make when the preset value drifts, even if this results in large deviations from the preset value.

Another disadvantage of open-loop systems is that they are poorly equipped to handle disturbances or changes in the conditions which may reduce its ability to complete the desired task. For example, the dryer door opens and heat is lost. The timing controller continues regardless for the full 30 minutes but the clothes are not heated or dried at the end of the drying process. This is because there is no information fed back to maintain a constant temperature.

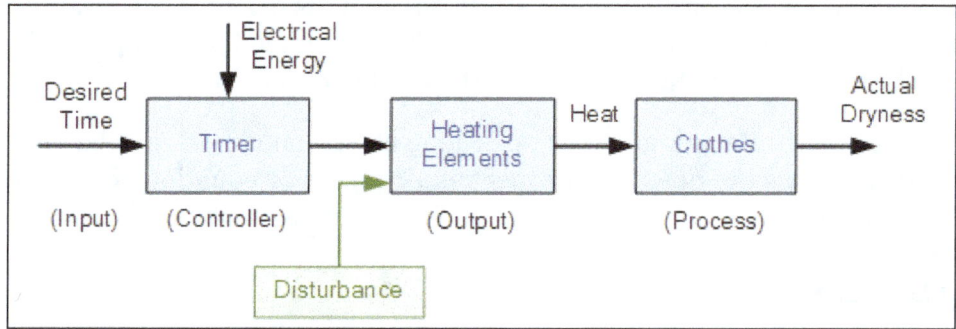

Then we can see that open-loop system errors can disturb the drying process and therefore requires extra supervisory attention of a user (operator). The problem with this anticipatory control approach is that the user would need to look at the process temperature frequently and take any corrective control action whenever the drying process deviated from its desired value of drying the clothes. This type of manual open-loop control which reacts before an error actually occurs is called feed forward control.

The objective of feed forward control, also known as predictive control, is to measure or predict any potential open-loop disturbances and compensate for them manually before the controlled variable deviates too far from the original set point. So for our simple example above, if the dryers door was open it would be detected and closed allowing the drying process to continue.

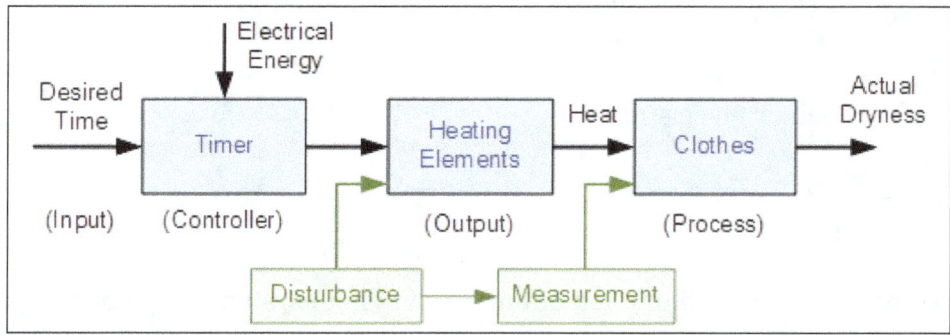

If applied correctly, the deviation from wet clothes to dry clothes at the end of the 30 minutes would be minimal if the user responded to the error situation (door open) very quickly. However, this feed forward approach may not be completely accurate if the system changes, for example the drop in drying temperature was not noticed during the 30 minute process.

Then we can define the main characteristics of an "Open-loop system" as being:

- There is no comparison between actual and desired values.

- An open-loop system has no self-regulation or control action over the output value.

- Each input setting determines a fixed operating position for the controller.

- Changes or disturbances in external conditions does not result in a direct output change (unless the controller setting is altered manually).

Any open-loop system can be represented as multiple cascaded blocks in series or a single block diagram with an input and output. The block diagram of an open-loop system shows that the signal path from input to output represents a linear path with no feedback loop and for any type of control system the input is given the designation θi and the output θo.

Generally, we do not have to manipulate the open-loop block diagram to calculate its actual transfer function. We can just write down the proper relationships or equations from each block diagram, and then calculate the final transfer function from these equations as shown.

Open-loop System

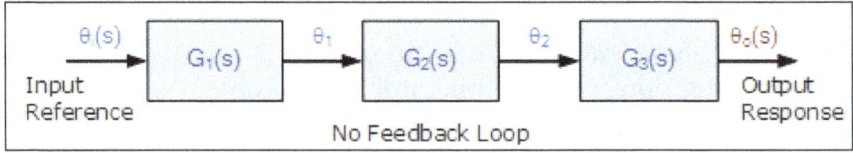

The Transfer Function of each block is therefore:

$$G_1 = \frac{\theta_1}{\theta_i}, \; G_2 = \frac{\theta_2}{\theta_1}, \; G_3 = \frac{\theta_0}{\theta_2},$$

The overall transfer function is given as:

$$G_1 \times G_2 \times G_3 = \frac{\theta_1}{\theta_i} \times \frac{\theta_2}{\theta_1} \times \frac{\theta_0}{\theta_2} = \frac{\theta_0}{\theta_i}$$

Then the Open-loop Gain is given simply as:

$$\text{Gain}, (G) = \frac{\theta_0(s)}{\theta_i(s)}$$

When G represents the Transfer Function of the system or subsystem, it can be rewritten as: G(s) = θo(s)/θi(s)

Open-loop control systems are often used with processes that require the sequencing of events with the aid of "ON-OFF" signals. For example a washing machines which requires the water to be switched "ON" and then when full is switched "OFF" followed by the heater element being switched "ON" to heat the water and then at a suitable temperature is switched "OFF", and so on.

This type of "ON-OFF" open-loop control is suitable for systems where the changes in load occur slowly and the process is very slow acting, necessitating infrequent changes to the control action by an operator.

Open-loop Motor Control

So for example, assume the DC motor controller as shown. The speed of rotation of the motor will depend upon the voltage supplied to the amplifier (the controller) by the potentiometer. The value of the input voltage could be proportional to the position of the potentiometer.

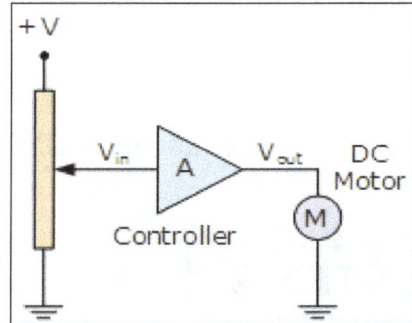

If the potentiometer is moved to the top of the resistance the maximum positive voltage will be supplied to the amplifier representing full speed. Likewise, if the potentiometer wiper is moved to the bottom of the resistance, zero voltage will be supplied representing a very slow speed or stop.

Then the position of the potentiometers slider represents the input, θi which is amplified by the amplifier (controller) to drive the DC motor (process) at a set speed N representing the output, θo of the system. The motor will continue to rotate at a fixed speed determined by the position of the potentiometer.

As the signal path from the input to the output is a direct path not forming part of any loop, the overall gain of the system will the cascaded values of the individual gains from the potentiometer, amplifier, motor and load. It is clearly desirable that the output speed of the motor should be identical to the position of the potentiometer, giving the overall gain of the system as unity.

However, the individual gains of the potentiometer, amplifier and motor may vary over time with changes in supply voltage or temperature, or the motors load may increase representing external disturbances to the open-loop motor control system.

But the user will eventually become aware of the change in the systems performance (change in motor speed) and may correct it by increasing or decreasing the potentiometers input signal accordingly to maintain the original or desired speed.

The advantages of this type of "open-loop motor control" is that it is potentially cheap and simple to implement making it ideal for use in well-defined systems were the relationship between input and output is direct and not influenced by any outside disturbances. Unfortunately this type of open-loop system is inadequate as variations or disturbances in the system affect the speed of the motor. Then another form of control is required.

Advantages

- Simple in construction and design:

- Economical.

- Easy to maintain.

- Generally stable.

- Convenient to use as output is difficult to measure.

Disadvantages

- They are inaccurate.

- They are unreliable.

- Any change in output cannot be corrected automatically.

CLOSED-LOOP SYSTEMS

A Closed-loop Control System, also known as a *feedback control system* is a control system which uses the concept of an open loop system as its forward path but has one or more feedback loops (hence its name) or paths between its output and its input. The reference to "feedback", simply means that some portion of the output is returned "back" to the input to form part of the systems excitation.

Closed-loop systems are designed to automatically achieve and maintain the desired output condition by comparing it with the actual condition. It does this by generating an error signal which is the difference between the output and the reference input. In other words, a "closed-loop system" is a fully automatic control system in which its control action being dependent on the output in some way.

Suppose we used a sensor or transducer (input device) to continually monitor the temperature or dryness of the clothes and feed a signal relating to the dryness back to the controller as shown below.

Closed-loop Control

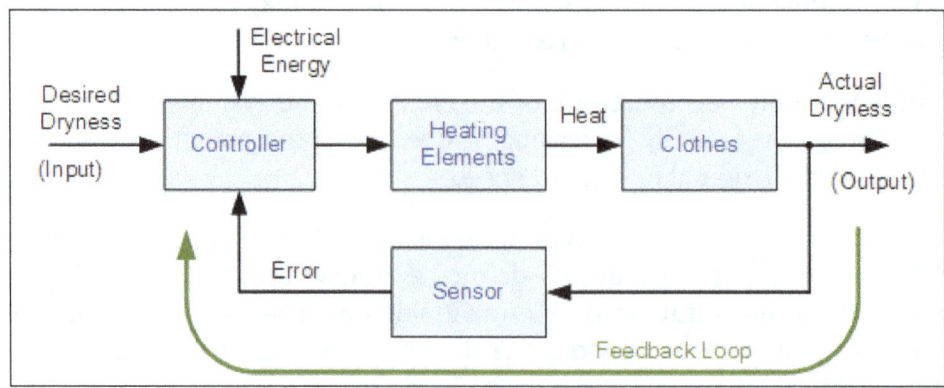

This sensor would monitor the actual dryness of the clothes and compare it with (or subtract it from) the input reference. The error signal (error = required dryness – actual dryness) is amplified by the controller, and the controller output makes the necessary correction to the heating system to reduce any error. For example if the clothes are too wet the controller may increase the temperature or drying time. Likewise, if the clothes are nearly dry it may reduce the temperature or stop the process so as not to overheat or burn the clothes, etc.

Then the closed-loop configuration is characterised by the feedback signal, derived from the sensor in our clothes drying system. The magnitude and polarity of the resulting error signal, would be directly related to the difference between the required dryness and actual dryness of the clothes.

Also, because a closed-loop system has some knowledge of the output condition, (via the sensor) it is better equipped to handle any system disturbances or changes in the conditions which may reduce its ability to complete the desired task.

For example, as before, the dryer door opens and heat is lost. This time the deviation in temperature is detected by the feedback sensor and the controller self-corrects the error to maintain a constant temperature within the limits of the preset value. Or possibly stops the process and activates an alarm to inform the operator.

As we can see, in a closed-loop control system the error signal, which is the difference between the input signal and the feedback signal (which may be the output signal itself or a function of the output signal), is fed to the controller so as to reduce the systems error and bring the output of the system back to a desired value. In our case the dryness of the clothes. Clearly, when the error is zero the clothes are dry.

The term Closed-loop control always implies the use of a feedback control action in order to reduce any errors within the system, and its "feedback" which distinguishes the main differences between an open-loop and a closed-loop system. The accuracy of the output thus depends on the feedback path, which in general can be made very accurate and within electronic control systems and circuits, feedback control is more commonly used than open-loop or feed forward control.

Closed-loop systems have many advantages over open-loop systems. The primary advantage of a closed-loop feedback control system is its ability to reduce a system's sensitivity to external disturbances, for example opening of the dryer door, giving the system a more robust control as any changes in the feedback signal will result in compensation by the controller.

Then we can define the main characteristics of Closed-loop Control as being:

- To reduce errors by automatically adjusting the systems input.

- To improve stability of an unstable system.

- To increase or reduce the systems sensitivity.

- To enhance robustness against external disturbances to the process.

- To produce a reliable and repeatable performance.

Whilst a good closed-loop system can have many advantages over an open-loop control system, its main disadvantage is that in order to provide the required amount of control, a closed-loop system must be more complex by having one or more feedback paths. Also, if the gain of the controller is too sensitive to changes in its input commands or signals it can become unstable and start to oscillate as the controller tries to over-correct itself, and eventually something would break. So we need to "tell" the system how we want it to behave within some pre-defined limits.

Closed-loop Summing Points

For a closed-loop feedback system to regulate any control signal, it must first determine the error between the actual output and the desired output. This is achieved using a summing point, also referred to as a comparison element, between the feedback loop and the systems input. These summing points compare a systems set point to the actual value and produce a positive or negative error signal which the controller responds too. Where: Error = Set point − Actual.

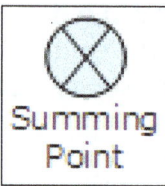

The symbol used to represent a summing point in closed-loop systems block-diagram is that of a circle with two crossed lines as shown. The summing point can either add signals together in which a Plus (+) symbol is used showing the device to be a "summer" (used for positive feedback), or it can subtract signals from each other in which case a Minus (−) symbol is used showing that the device is a "comparator" (used for negative feedback) as shown.

Summing Point Types

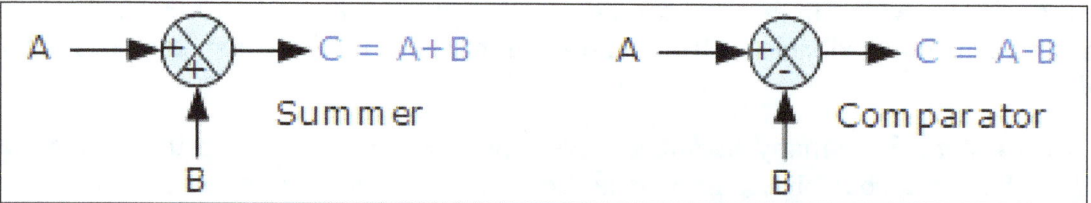

Note that summing points can have more than one signal as inputs either adding or subtracting but only one output which is the algebraic sum of the inputs. Also the arrows indicate the direction of the signals. Summing points can be cascaded together to allow for more input variables to be summed at a given point.

Closed-loop System Transfer Function

The Transfer Function of any electrical or electronic control system is the mathematical relationship between the systems input and its output, and hence describes the behaviour of the system. Note also that the ratio of the output of a particular device to its input represents its gain. Then we can correctly say that the output is always the transfer function of the system times the input. Consider the closed-loop system below.

Typical Closed-loop System Representation

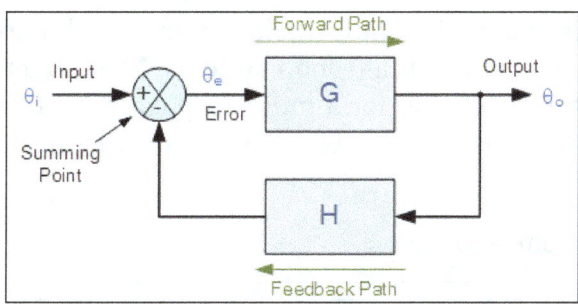

Where: block G represents the open-loop gains of the controller or system and is the forward path, and block H represents the gain of the sensor, transducer or measurement system in the feedback path.

To find the transfer function of the closed-loop system above, we must first calculate the output signal θ_o in terms of the input signal θ_i. To do so, we can easily write the equations of the given block-diagram as follows.

The output from the system is equal to: Output = G x Error

Note that the error signal, θ_e is also the input to the feed-forward block: G

The output from the summing point is equal to: Error = Input - H x Output

If H = 1 (unity feedback) then:

The output from the summing point will be: Error (θ_e) = Input - Output

Eliminating the error term, then:

The output is equal to: Output = G x (Input - H x Output)

Therefore: G x Input = Output + G x H x Output

Rearranging the above gives us the closed-loop transfer function of:

$$\frac{\text{Output}}{\text{Input}} = \frac{\theta_0}{\theta_i} = \frac{G}{1 + GH}$$

The above equation for the transfer function of a closed-loop system shows a Plus (+) sign in the denominator representing negative feedback. With a positive feedback system, the denominator will have a Minus (–) sign and the equation becomes: 1 - GH.

We can see that when H = 1 (unity feedback) and G is very large, the transfer function approaches unity as:

$$\frac{\text{Output}}{\text{Input}} \rightarrow 1$$

Also, as the systems steady state gain G decreases, the expression of: G/(1 + G) decreases much more slowly. In other words, the system is fairly insensitive to variations in the systems gain represented by G, and which is one of the main advantages of a closed-loop system.

Multi-loop Closed-loop System

Whilst our example above is of a single input, single output closed-loop system, the basic transfer function still applies to more complex multi-loop systems. Most practical feedback circuits have some form of multiple loop control, and for a multi-loop configuration the transfer function between a controlled and a manipulated variable depends on whether the other feedback control loops are open or closed.

Consider the multi-loop system below.

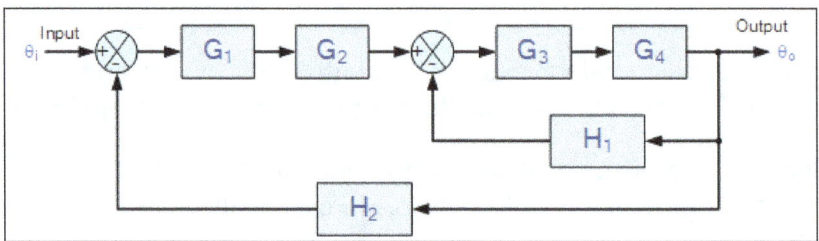

Any cascaded blocks such as G_1 and G_2 can be reduced, as well as the transfer function of the inner loop as shown.

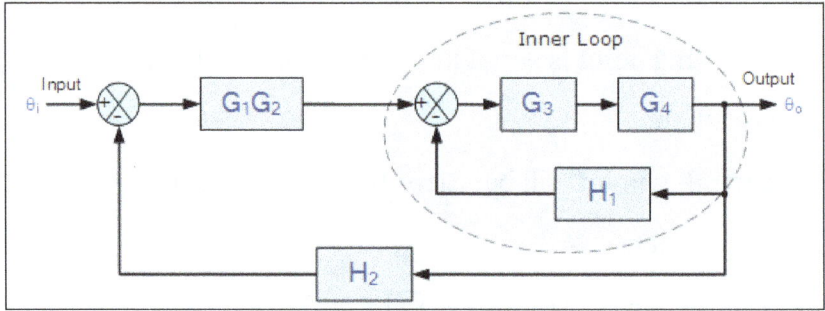

After further reduction of the blocks we end up with a final block diagram which resembles that of the previous single-loop closed-loop system.

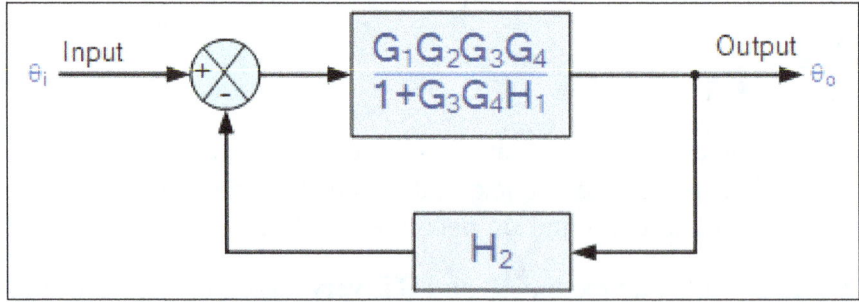

And the transfer function of this multi-loop system becomes:

$$\frac{\text{Output}}{\text{Input}} = \frac{\theta_0}{\theta_i} = \frac{G_1\,G_2\,G_3\,G_4}{1 + G_3\,G_4\,H_1 + G_1\,G_2\,G_3\,G_4\,H_2}$$

Then we can see that even complex multi-block or multi-loop block diagrams can be reduced to give one single block diagram with one common system transfer function.

Closed-loop Motor Control

So how can we use Closed-loop Systems in Electronics. If we connected a speed measuring transducer, such as a tachometer to the shaft of the DC motor, we could detect its speed and send a signal proportional to the motor speed back to the amplifier. A tachometer, also known as a tacho-generator is simply a permanent-magnet DC generator which gives a DC output voltage proportional to the speed of the motor.

Then the position of the potentiometers slider represents the input, θ_i which is amplified by the amplifier (controller) to drive the DC motor at a set speed N representing the output, θ_o of the system, and the tachometer T would be the closed-loop back to the controller. The difference between the input voltage setting and the feedback voltage level gives the error signal as shown.

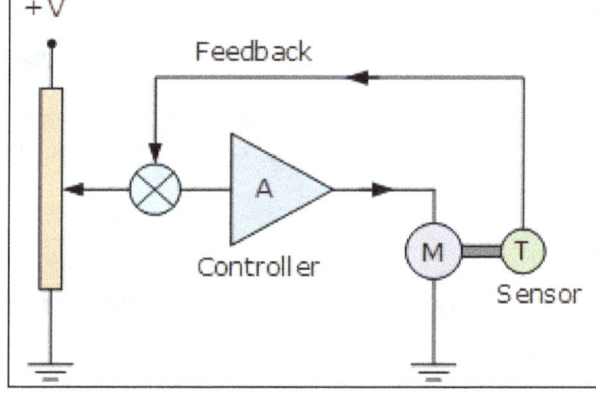

Closed-loop Motor Control.

Any external disturbances to the closed-loop motor control system such as the motors load increasing would create a difference in the actual motor speed and the potentiometer input set point.

This difference would produce an error signal which the controller would automatically respond too adjusting the motors speed. Then the controller works to minimize the error signal, with zero error indicating actual speed which equals set point.

Electronically, we could implement such a simple closed-loop tachometer-feedback motor control circuit using an operational amplifier (op-amp) for the controller as shown.

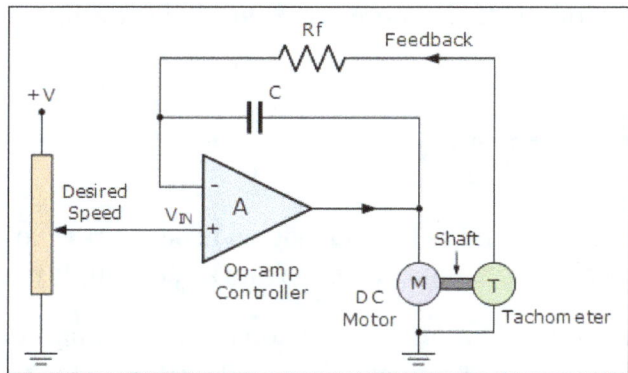

Closed-loop Motor Controller Circuit.

This simple closed-loop motor controller can be represented as a block diagram as shown.

Block Diagram for the Feedback Controller

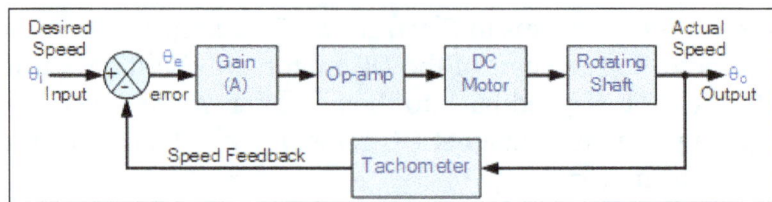

A closed-loop motor controller is a common means of maintaining a desired motor speed under varying load conditions by changing the average voltage applied to the input from the controller. The tachometer could be replaced by an optical encoder or Hall-effect type positional or rotary sensor.

Advantages

- Closed loop control systems are more accurate even in the presence of non-linearity.

- Highly accurate as any error arising is corrected due to presence of feedback signal.

- Bandwidth range is large.

- Facilitates automation.

- The sensitivity of system may be made small to make system more stable.

- This system is less affected by noise.

Disadvantages

- They are costlier.

- They are complicated to design.

- Required more maintenance.

- Feedback leads to oscillatory response.

- Overall gain is reduced due to presence of feedback.

- Stability is the major problem and more care is needed to design a stable closed loop system.

FEEDBACK SYSTEMS

Feedback Systems process signals and as such are signal processors. The processing part of a feedback system may be electrical or electronic, ranging from a very simple to a highly complex circuits.

Simple analogue feedback control circuits can be constructed using individual or discrete components, such as transistors, resistors and capacitors, etc, or by using microprocessor-based and integrated circuits (IC's) to form more complex digital feedback systems.

As we have seen, open-loop systems are just that, open ended, and no attempt is made to compensate for changes in circuit conditions or changes in load conditions due to variations in circuit parameters, such as gain and stability, temperature, supply voltage variations and external disturbances. But the effects of these "open-loop" variations can be eliminated or at least considerably reduced by the introduction of Feedback.

A feedback system is one in which the output signal is sampled and then fed back to the input to form an error signal that drives the system.

Feedback Systems are very useful and widely used in amplifier circuits, oscillators, process control systems as well as other types of electronic systems. But for feedback to be an effective tool it must be controlled as an uncontrolled system will either oscillate or fail to function. The basic model of a feedback system is given as:

Feedback System Block Diagram Model

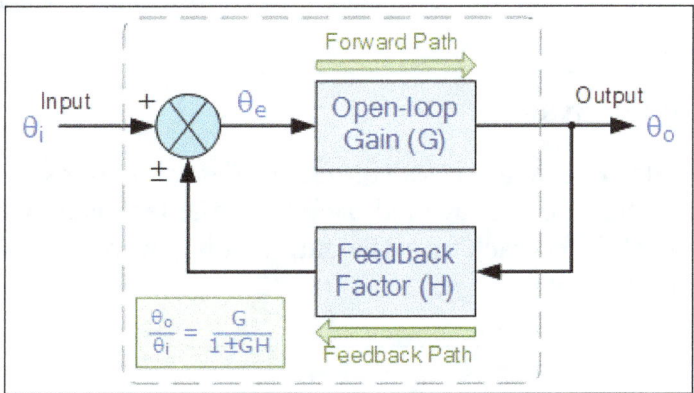

This basic feedback loop of sensing, controlling and actuation is the main concept behind a feedback control system and there are several good reasons why feedback is applied and used in electronic circuits:

- Circuit characteristics such as the systems gain and response can be precisely controlled.

- Circuit characteristics can be made independent of operating conditions such as supply voltages or temperature variations.

- Signal distortion due to the non-linear nature of the components used can be greatly reduced.

- The Frequency Response, Gain and Bandwidth of a circuit or system can be easily controlled to within tight limits.

Classification of Feedback Systems

Thus far we have seen the way in which the output signal is "fed back" to the input terminal, and for feedback systems this can be of either, Positive Feedback or Negative Feedback. But the manner in which the output signal is measured and introduced into the input circuit can be very different leading to four basic classifications of feedback.

Based on the input quantity being amplified, and on the desired output condition, the input and output variables can be modelled as either a voltage or a current. As a result, there are four basic classifications of single-loop feedback system in which the output signal is fed back to the input and these are:

- Series-Shunt Configuration – Voltage in and Voltage out or Voltage Controlled Voltage Source (VCVS).

- Shunt-Shunt Configuration – Current in and Voltage out or Current Controlled Voltage Source (CCVS).

- Series-Series Configuration – Voltage in and Current out or Voltage Controlled Current Source (VCCS).

- Shunt-Series Configuration – Current in and Current out or Current Controlled Current Source (CCCS).

These names come from the way that the feedback network connects between the input and output stages as shown.

Series-Shunt Feedback Systems

Series-Shunt Feedback, also known as *series voltage feedback*, operates as a voltage-voltage controlled feedback system. The error voltage fed back from the feedback network is in *series* with the input. The voltage which is fed back from the output being proportional to the output voltage, V_o as it is parallel, or shunt connected.

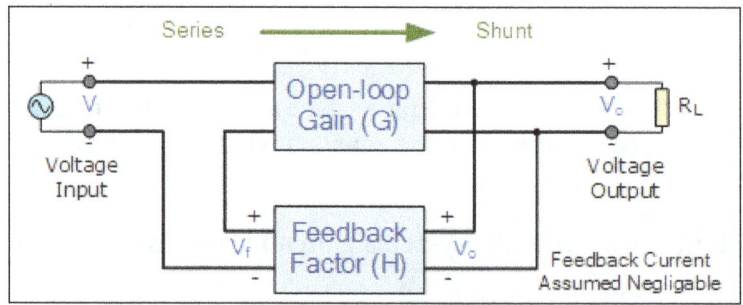

Series-Shunt Feedback System.

For the series-shunt connection, the configuration is defined as the output voltage, V_{out} to the input voltage, V_{in}. Most inverting and non-inverting operational amplifier circuits operate with series-shunt feedback producing what is known as a "voltage amplifier". As a voltage amplifier the ideal input resistance, R_{in} is very large, and the ideal output resistance, R_{out} is very small.

Then the "series-shunt feedback configuration" works as a true voltage amplifier as the input signal is a voltage and the output signal is a voltage, so the transfer gain is given as: Av = Vout ÷ Vin. Note that this quantity is dimensionless as its units are volts/volts.

Shunt-Series Feedback Systems

Shunt-Series Feedback, also known as *shunt current feedback*, operates as a current-current controlled feedback system. The feedback signal is proportional to the output current, I_o flowing in the load. The feedback signal is fed back in parallel or *shunt* with the input as shown.

For the shunt-series connection, the configuration is defined as the output current, I_{out} to the input current, I_{in}. In the shunt-series feedback configuration the signal fed back is in parallel with the input signal and as such its the currents, not the voltages that add.

This parallel shunt feedback connection will not normally affect the voltage gain of the system, since for a voltage output a voltage input is required. Also, the series connection at the output increases output resistance, Rout while the shunt connection at the input decreases the input resistance, R_{in}.

Then the "shunt-series feedback configuration" works as a true current amplifier as the input signal is a current and the output signal is a current, so the transfer gain is given as: $A_i = I_{out} \div I_{in}$. Note that this quantity is dimensionless as its units are amperes/amperes.

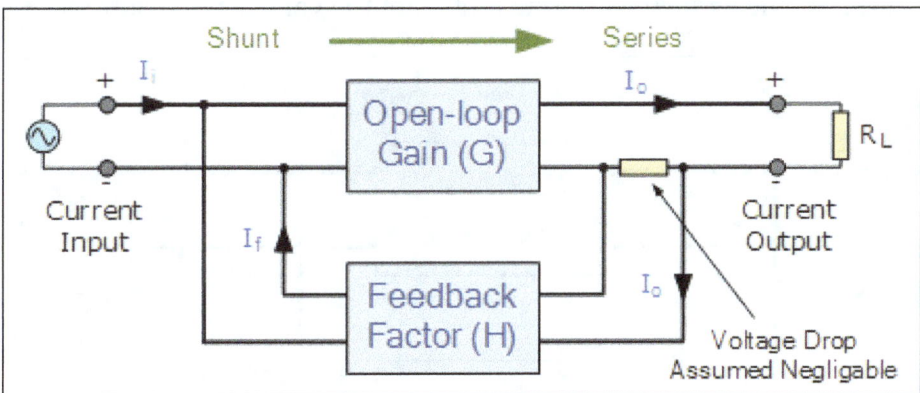

Series-Series Feedback Systems

Series-Series Feedback Systems, also known as *series current feedback*, operates as a voltage-current controlled feedback system. In the series current configuration the feedback error signal is in *series* with the input and is proportional to the load current, I_{out}. Actually, this type of feedback converts the current signal into a voltage which is actually fed back and it is this voltage which is subtracted from the input.

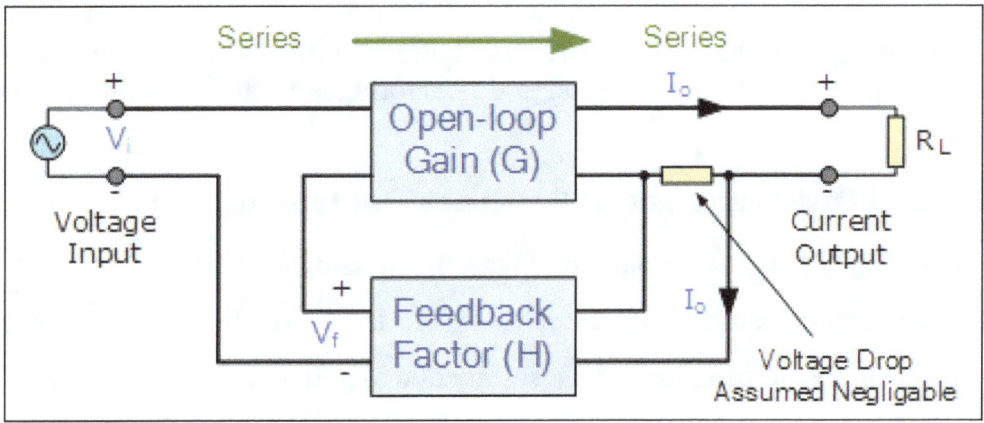

For the series-series connection, the configuration is defined as the output current, I_{out} to the input voltage, V_{in}. Because the output current, I_{out} of the series connection is fed back as a voltage, this increases both the input and output impedances of the system. Therefore, the circuit works best as a transconductance amplifier with the ideal input resistance, R_{in} being very large, and the ideal output resistance, R_{out} is also very large.

Then the "series-series feedback configuration" functions as transconductance type amplifier system as the input signal is a voltage and the output signal is a current. then for a series-series feed-back circuit the transfer gain is given as: $G_m = I_{out} \div V_{in}$.

Shunt-Shunt Feedback Systems

Shunt-Shunt Feedback Systems, also known as *shunt voltage feedback*, operates as a current-voltage controlled feedback system. In the shunt-shunt feedback configuration the signal fed back is in parallel with the input signal. The output voltage is sensed and the current is subtracted from the input current in shunt, and as such its the currents, not the voltages that subtract.

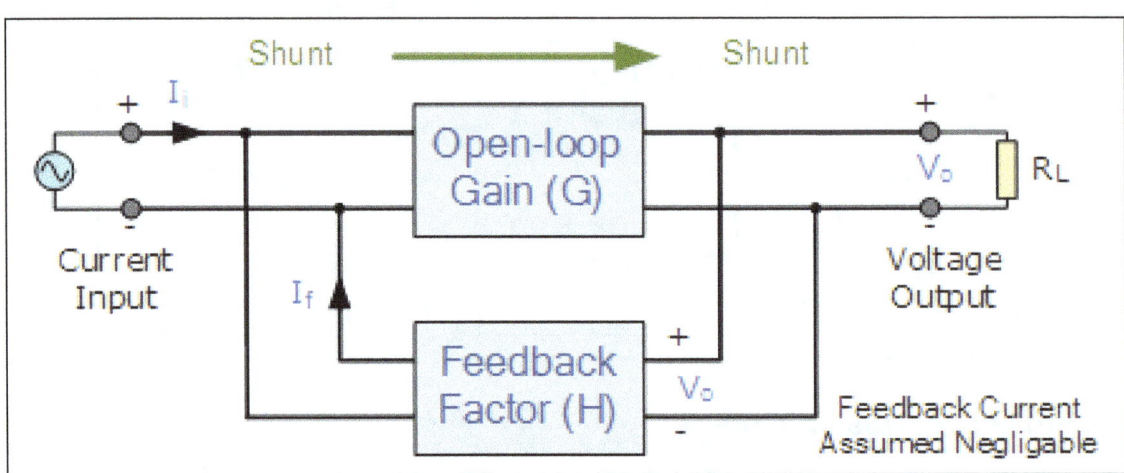

For the shunt-shunt connection, the configuration is defined as the output voltage, V_{out} to the input current, I_{in}. As the output voltage is fed back as a current to a current-driven input port, the shunt connections at both the input and output terminals reduce the input and output impedance. therefore the system works best as a transresistance system with the ideal input resistance, R_{in} being very small, and the ideal output resistance, R_{out} also being very small.

Then the shunt voltage configuration works as transresistance type voltage amplifier as the input signal is a current and the output signal is a voltage, so the transfer gain is given as: $R_m = V_{out} \div I_{in}$.

Advantages and Disadvantages of Feedback Control Systems

The following advantages are the fundamental reasons for using feedback.

- Many unnecessary disturbances and noise signals from outside the system can be rejected.

- The change in the performance of the system due to parameter variations is reduced.

- The steady-state error of the system can be relatively small.

- The transient behavior of the process can be easily manipulated.

- The feedback is compared with the desired state in order to take corrective measures.

On the other hand, using feedback can have several disadvantages:

- The system is complicated by the increased number of components, such as sensors and error detectors.

- The overall gain of the system is reduced and must be compensated for in the design.

- The system may not be stable (it may oscillate or depart greatly from the desired output), even though the comparable open-loop system is stable.

- The error detector is necessary in order to compare two states.

- If there is a change in an Output, it will affect the system input.

Because the advantages of feedback exceed its disadvantages in most cases, feedback has become the major concept in the design of control system.

NEGATIVE FEEDBACK SYSTEMS

Feedback is the process by which a fraction of the output signal, either a voltage or a current, is used as an input. If this feed back fraction is opposite in value or phase ("anti-phase") to the input signal, then the feedback is said to be Negative Feedback, or *degenerative feedback*.

Negative feedback opposes or subtracts from the input signals giving it many advantages in the design and stabilisation of control systems. For example, if the systems output changes for any reason, then negative feedback affects the input in such a way as to counteract the change.

Feedback reduces the overall gain of a system with the degree of reduction being related to the systems open-loop gain. Negative feedback also has effects of reducing distortion, noise, sensitivity to external changes as well as improving system bandwidth and input and output impedances.

Feedback in an electronic system, whether negative feedback or positive feedback is unilateral in direction. Meaning that its signals flow one way only from the output to the input of the system. This then makes the loop gain, G of the system independent of the load and source impedances.

As feedback implies a closed-loop system it must therefore have a summing point. In a negative feedback system this summing point or junction at its input subtracts the feedback signal from the input signal to form an error signal, β which drives the system. If the system has a positive gain, the feedback signal must be subtracted from the input signal in order for the feedback to be negative as shown.

Negative Feedback Circuit

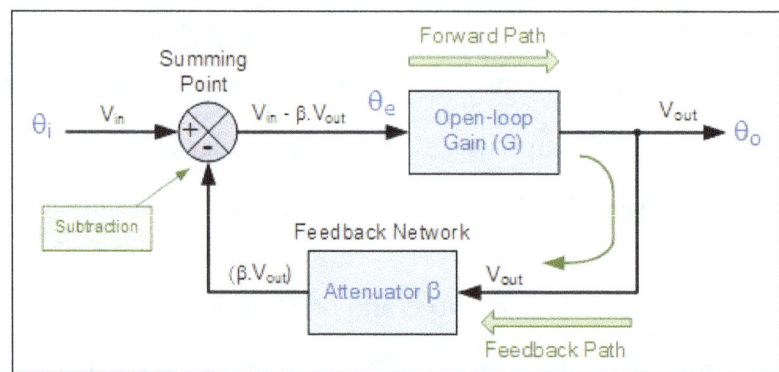

The circuit represents a system with positive gain, G and feedback, β. The summing junction at its input subtracts the feedback signal from the input signal to form the error signal V_{in} - βG, which drives the system.

Then using the basic closed-loop circuit above we can derive the general feedback equation as being:

Negative Feedback Equation

$$\text{System Gain, } G = \frac{V_{out}}{V_{in}} \quad G = \text{open loop voltage gain}$$

$$G \times V_{in} = V_{out}$$

$$G(V_{in} - \beta V_{out}) = V_{out} \quad \beta \text{ is the feedback fraction}$$

$$G.V_{in} - \beta.G.V_{out} = V_{out} \quad \beta G = \text{the loop gain}$$

$$G.V_{in} = V_{out}(1 + \beta G) \quad 1 + \beta G = \text{the feedback factor}$$

$$\therefore \frac{V_{out}}{V_{in}} = Gv = \frac{G}{1 + \beta G} \quad Gv = \text{closed loop voltage gain}$$

We see that the effect of the negative feedback is to reduce the gain by the factor of: 1 + βG. This factor is called the "feedback factor" or "amount of feedback" and is often specified in decibels (dB) by the relationship of 20 log (1+ βG).

Effects of Negative Feedback

If the open-loop gain, G is very large, then βG will be much greater than 1, so that the overall gain of the system is roughly equal to 1/β. If the open-loop gain decreases due to frequency or the effects of system ageing, providing that βG is still relatively large, the overall system gain does not change very much. So negative feedback tends to reduce the effects of gain change giving what is generally called "gain stability".

Negative Feedback Example

A system has a gain of 80dB without feedback. If the negative feedback fraction is 1/50th.

Calculate the closed-loop gain of the system in dB with the addition of negative feedback.

$$80\,dB = 20\log(G)$$
$$\therefore G = \text{antilog}\,10^{80/20} = 10,000$$

$$Gv = \frac{G}{1+G\beta} = \frac{10,000}{1+10,000.\left(\dfrac{1}{50}\right)} = 49.75$$

$$\therefore Gv_{(dB)} = 20\log(49.75) = 34\,dB$$

Then we can see that the system has a loop gain of 10,000 and a closed-loop gain of 34dB.

If after 5 years the loop gain of the system without negative feedback has fallen to 60dB and the feedback fraction has remained constant at 1/50th. Calculate the new closed-loop gain value of the system.

$$60\,dB = 20\log(G)$$
$$\therefore G = \text{antilog}\,10^{60/20} = 1,000$$

$$Gv = \frac{G}{1+G\beta} = \frac{1,000}{1+1,000.\left(\dfrac{1}{50}\right)} = 47.6$$

$$\therefore Gv_{(dB)} = 20\log(47.6) = 33.5\,dB$$

Then we can see from the two examples that without feedback, after 5 years of use the systems gain has fallen from 80dB down to 60dB, (10,000 to 1,000) a drop in open loop gain of about 25%.

However with the addition of negative feedback the systems gain has only fallen from 34dB to 33.5dB, a reduction of less than 1.5%, which proves that negative feedback gives added stability to a systems gain.

Therefore we can see that by applying negative feedback to a system greatly reduces its overall gain compared to its gain without feedback.

The systems gain without feedback can be very large but not precise as it may change from one system device to the next, then it is possible to design a system with sufficient open-loop gain that, after the negative feedback has been added, the overall gain matches the desired value.

Also, if the feedback network is constructed from passive elements having stable characteristics, the overall gain becomes very steady and unaffected by variation in the systems inherent open-loop gain.

Negative Feedback in Operational Amplifiers

Operational amplifiers (op-amps) are the most commonly used type of linear integrated circuit but they have a very high gain. The open-loop voltage gain, A_{VOL}, of a standard 741 op-amp is its voltage gain when there is no negative feedback applied and the open-loop voltage gain of an op-amp is the ratio of its output voltage, V_{out}, to its differential input voltage, V_{in}, (V_{out}/V_{in}).

The typical value of A_{VOL} for a 741 op-amp is more than 200,000 (106dB). So an input voltage signal of only 1mV, would result in an output voltage of over 200 volts! forcing the output immediately into saturation. Obviously this high open-loop voltage gain needs to be controlled in some way, and we can do just that by using negative feedback.

The use of negative feedback can significantly improve the performance of an operational amplifier and any op-amp circuit that does not use negative feedback is considered too unstable to be useful. But how can we use negative feedback to control an op-amp. Well consider the circuit below of a Non-inverting Operational Amplifier.

Non-inverting Op-amp Circuit

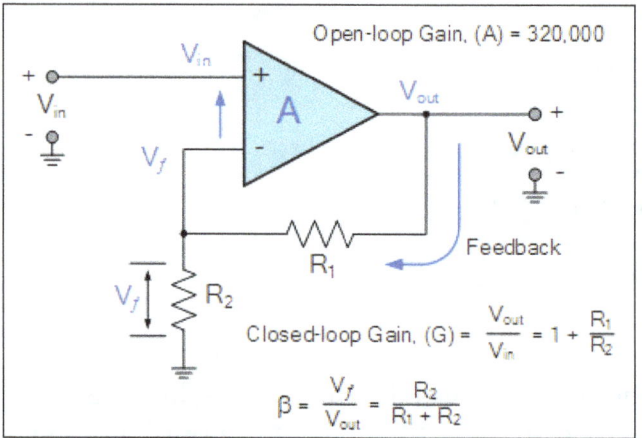

An operational amplifier with an open-loop voltage gain, A_{VOL} of 320,000 without feedback is to be used as a non-inverting amplifier. Calculate the values of the feedback resistances, R_1 and R_2 required to stabilise the circuit with a closed loop gain of 20.

The generalised closed-loop feedback equation we derived above is given as:

$$j = \frac{A}{1 + \beta A}$$

By rearranging the feedback formula we get a feedback fraction, β of:

$$j(1 + \beta A) = A$$

$$+ \beta A = \frac{A}{G}$$

$$j A = \frac{A}{G} - 1$$

$$\therefore \beta = \frac{1}{G} - \frac{1}{A}$$

Then putting the values of: A = 320,000 and G = 20, into the above equation we get the value of β as:

$$\beta = \frac{1}{G} - \frac{1}{A}$$

$$\beta = \frac{1}{20} - \frac{1}{320,000} = 0.05$$

Because in this case the open-loop gain of the op-amp is very high (A = 320,000), the feedback fraction, β will be roughly equal to the reciprocal of the closed-loop gain 1/G only as the value of 1/A will be incredibly small. Then β (the feedback fraction) is equal to 1/20 = 0.05.

As the resistors, R_1 and R_2 form a simple series-voltage potential divider network across the non-inverting amplifier, the closed-loop voltage gain of the circuit will be determined by the ratios of these resistances as:

$$\beta = \frac{V_f}{V_{OUT}} = \frac{R_2}{R_1 + R_2}$$

If we assume resistor R_2 has a value of 1,000Ω, or 1kΩ, then the value of resistor R_1 will be:

$$\beta = \frac{R_2}{R_1 + R_2}$$

$$\therefore R = \frac{R_2 - \beta R_2}{\beta}$$

$$= \frac{1000 - (0.05 \times 1000)}{0.05} = 19,000\Omega = 19\,K\Omega$$

Then for the non-inverting amplifier circuit about to have a closed-loop gain of 20, the values of the negative feedback resistors required will be in this case, R_1 = 19kΩ and R_2 = 1kΩ, giving us a non-inverting amplifier circuit of:

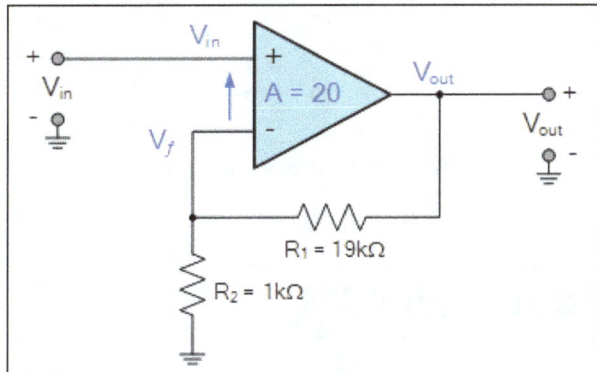

There are many advantages to using feedback within a systems design, but the main advantages of using Negative Feedback in amplifier circuits is to greatly improve their stability, better tolerance to component variations, stabilisation against DC drift as well as increasing the amplifiers bandwidth.

Examples of negative feedback in common amplifier circuits include the resistor R_f in op-amp circuits as we have seen above, resistor, R_S in FET based amplifiers and resistor, R_E in bipolar transistor (BJT) amplifiers.

Advantages and Disadvantages of Negative Feedback Systems

Advantages

- The negative feedback reduces noise.
- It has highly stabilized gain.
- It can control step response of amplifier.
- It has less harmonic distortion.
- It has less amplitude distortion.
- It has less phase distortion.
- Input and output impedances can be modified as desired.
- It can increase or decrease output impedances.
- It has higher fidelity i.e. more linear operation.
- It has less frequency distortion.

Disadvantages

- It Increase output resistance in case of current shunt and current series feedback amplifiers.
- Reduction in gain.

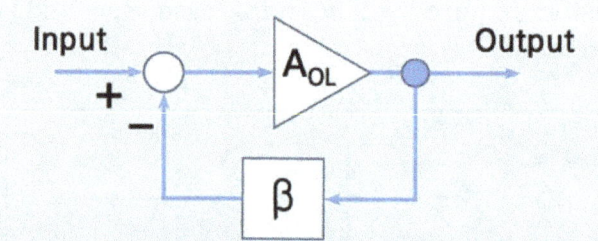

Negative feedback amplifier.

LOGIC CONTROL SYSTEM

The pneumatic cylinder moves in a linear dimension until it reaches the limit switch at the extended end. The cylinder is controlled with a simple two position, four-way solenoid valve as shown. The solenoid valve shown is activated by an electrical current passing through the solenoid coil. This type of simple ON/OFF programming has traditionally been done by relay control systems.

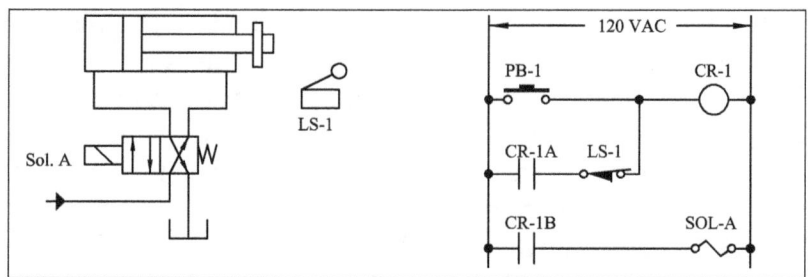

LC-: Simple pneumatic and logic control system.

A relay control system for the simple system of figure LC- is also shown below. This schematic figure represents a type of programming frequently referred to as "ladder logic" by industrial electricians. The two parts of a relay are both shown in the figure. Electrical relays (figure LC-) have a control circuit and one or more sets of outputs. The coil of the relay forms part of an electromagnet which activates a set of contacts (contacts similar to "points" in an pre-70's auto). Electrical current passing through the coil of the relay (the "control relay") closes one of these sets of contacts (CR-1B) which allows current to flow through the pneumatic valve solenoid, SOL-A. Another set of contacts, CR-1A in figure LC-, is used to "hold" the contacts closed once they have been energized, by providing an alternate path for electrical current through the control relay. A momentary contact push-button PB-1 (normally open or N.O.) is provided for initiating motion. When PB-1 is pressed, current flows through the actuating circuit of relay CR-1, which closes the output contacts (CR-1A and CR-1B). When PB-1 is released, these contacts remain closed due to electrical current path through the closed relay contacts CR-1A and the normally closed limit switch LS-1. Relay CR-1 remains energized until the limit switch LS-1 activated by the cylinder. Once this limit switch is activated, the current flow through the control relay CR-1 is interrupted, and the contacts CR-1A and CR-1B both open. The solenoid SOL-A is de-energized, therefore the spring shifts the solenoid back to the right position, which causes the cylinder to retract. The circuit is inactive until a subsequent pressing of the push-button PB-1.

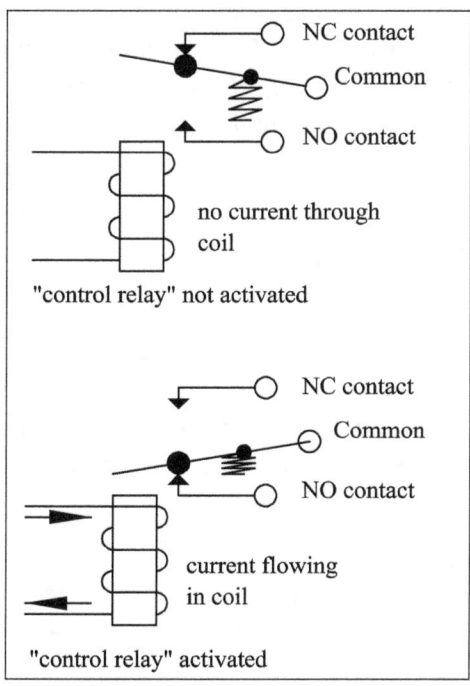

LC-: Control relay with NO and NC contacts.

Figure LC- shown below the most common components of ladder logic diagrams. Input elements include limit switches, momentary contact push-buttons, pressure switches, manual switches, and relay contacts. Typical outputs include solenoid coils, control relay coils, pilot lights, and annunciators (or horns). Note that each of the inputs is available in both normally open (NO) and normally closed (NC) configurations. This distinction is easily explained by observing the limit switch configurations. A normally closed limit switch will carry current if it is not activated (the "normal" state). If a normally closed limit switch is pressed, then it no longer will carry current. A normally open limit switch is the opposite - it will not carry current inactivated, it must be pressed to allow current to flow through it.

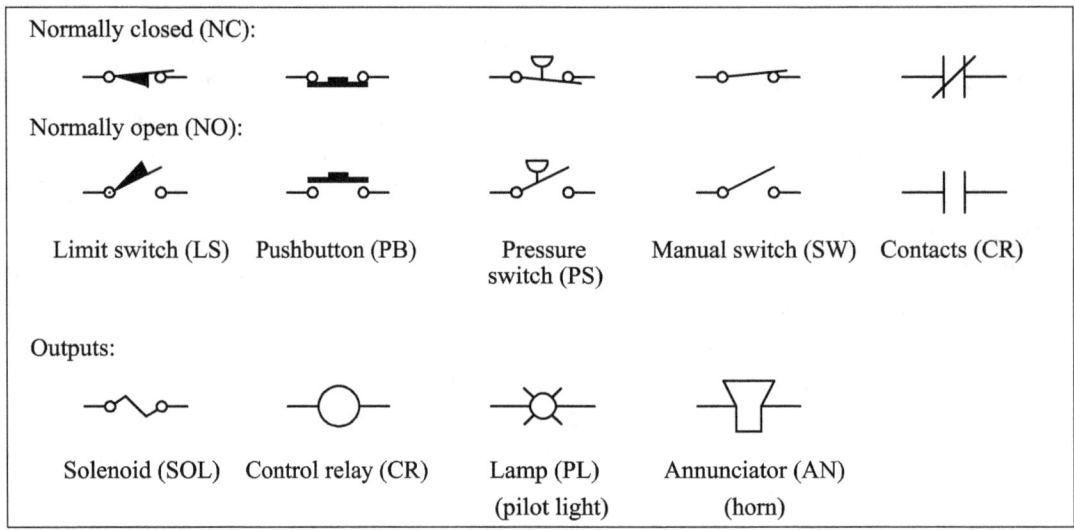

LC-: Ladder logic schematic elements.

Several simple ladder logic diagrams are given below to illustrate the use of ladder logic. In each of these diagrams an "equivalent" BASIC statement is also given for comparison (all variables considered to be logical variables). In the first diagram, a single momentary contact push-button activates a control relay coil. The control relay CR-1 is considered the output and the momentary contact pushbutton PB-1 is the input. Note that the coil is activated only while the push-button is held down, once it is released the coil is no longer activated.

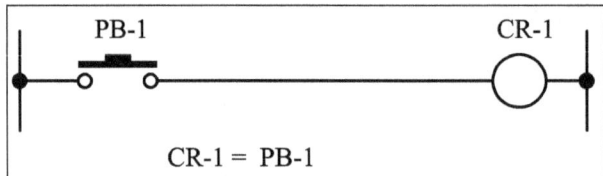

In the next example a logical "AND" configuration is implemented. Two inputs (PB-1 and LS-2) are used to activate a single output (SOL-2). In order to activate the output, both of the input must be activated at the same time. Once the push-button PB-1 is released or the limit switch LS-1 is not activated, the solenoid coil SOL-2 is no longer active.

A logical "OR" configuration is shown in this diagram. If either of the two inputs (LS-1 or LS2) is activated, or both of them, then the output pilot light PL-3 will come on. If neither inputs is activated, then the output PL-3 will be off. Note that two alternate paths for electrical current are provided through the two limit switches.

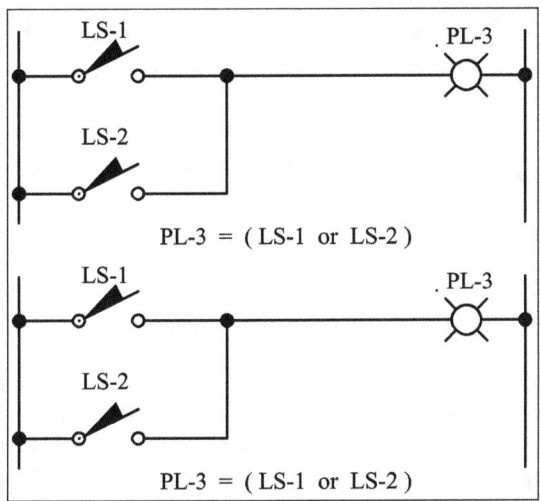

A combination of logical "AND" and "OR" is given in this example. The contacts CR-3A must be closed and at least one of the other two inputs (LS-1 and PS-2) must be activated. Note that the CR-3A contacts are closed by activating the control relay CR-3 on another rung of the ladder diagram. Contacts associated with CR-3 can appear many times on the ladder diagram, but the control relay itself can appear as an output on only one rung.

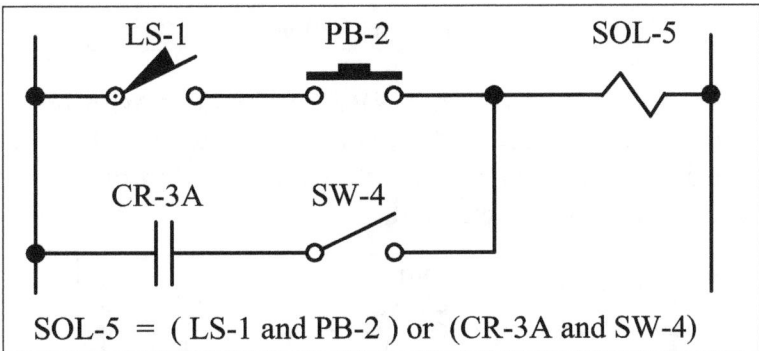

Another combination of logical "AND": and logical "OR" elements is given in this example. There are two ways to activate the output SOL-5, activate LS-1 and PB-2 simultaneously, or activate contacts CR-3A and switch SW-4 simultaneously. Momentary contact push-buttons are much more common than switches (like SW-4), due to safety considerations. Again, note the two alternate paths for current to flow from the left to the right sides of the "ladder."

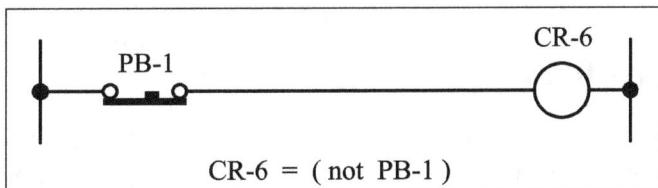

The following examples show the use of normally closed contacts. In the first example, the control relay coil CR-6 is activated as long as the push-button PB-1 is not pressed. Pressing PB-1 will de-activate the coil CR-6.

A logical "AND" function is shown in this example along with normally closed elements. As long as neither of the inputs (PB-1 and LS-2) is activated, the output solenoid coil SOL-7 will be activated. If either input is activated, then the output solenoid is de-energized.

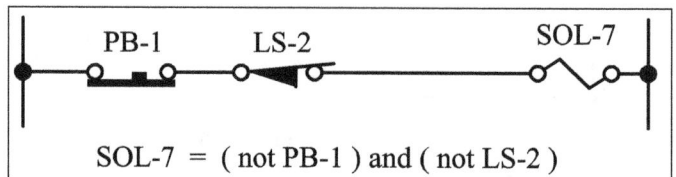

A combination "AND" and "OR" rung is shown with normally closed elements. The output solenoid SOL-8 will be activated as long as CR-2 is not activated and either LS-1 or PS-3 is not activated. Note that the normally closed contacts CR-2A remain closed until the control relay coil CR-2 is activated on another rung of the ladder diagram. Normally closed contacts (like CR-2A) can occur as many times as needed in the ladder diagram.

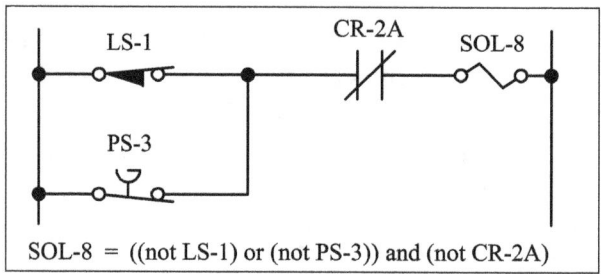

The final example shows a logical "AND" and "OR" combination with both normally open and normally closed elements. The output horn AN9 will be activated if either (or both) of the following conditions is satisfied: input LS-1 is not activated and push-button PB-1 is, or switch SW-3 is activated and control relay CR-4 is not activated.

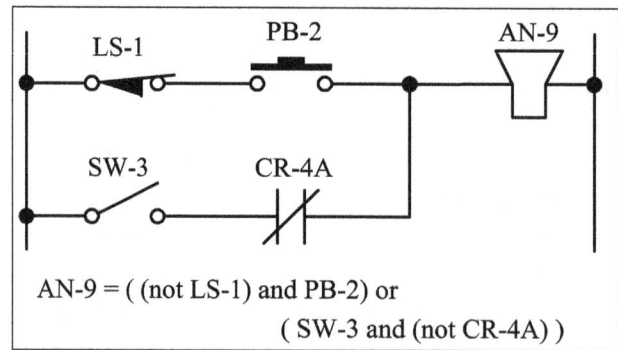

Programmable Controllers

One of the disadvantages of the relay logic systems discussed previously is the difficult nature of the "programming." The program logic is "hard-wired" by the interconnection of the relays, limit switches, timers, counters, etc. Changing the task performed by the simple system of figure LC-requires physically moving the wires from the relays and limit switches and placing them in the desired new configuration. For circuits with only three or four components this is not difficult. However, systems containing ten to several hundred individual components are not uncommon in industrial automation systems.

The programmable logic controller (PLC) was developed in the early 60's to overcome the deficiencies of relay logic systems. Programmable logic is implemented using a microcomputer instead of the hard-wired logic of the conventional hard-wired relay system. The major advantage of PLC's (frequently referred to as just programmable controllers or PC's) is that the programming can be done in ladder logic, just like relay logic systems. Electricians and technicians can readily adapt to this familiar type of programming. A computer language like BASIC or Pascal might be too intimidating and is not required to implement straightforward machine logic.

The major criteria for specifying PLC's are the number of input contacts that can be read and the number of output switches that can be controlled. Small PLC's might have 8 to 12 inputs and outputs, while larger models can use 100 or more I/O (input/ output) points. Inputs are usually 0-120 volts AC or 0-24 volts DC. Output options frequently include relay contacts, triac (120 VAC) or 24 volt (open collector). Some of the newer PLC models have such advanced features as analog inputs (0 - 10 Volts), PID (proportional- integral-derivative) control loops, and serial (RS-232) communications capabilities.

Figure LC- shows a programmable controller ladder diagram for the same simple system of figure LC-1. The "internal" contact labeled X1 is connected to the input push-button, PB-1. The normally open limit switch, LS-1, is wired to the input contact X2. The internal contact C5 replaces the control relay CR-1 and its two pairs of contacts. The output solenoid coil, SOL-A, is connected to the output contact Y2. By comparing this figure to the relay system of figure LC-1, the similarities between PLC programming and relay logic is obvious. One simplifying difference is that internal registers (such as X1, X2, and C5) can be used as replacements for inputs and control relay circuits. An essentially unlimited number of input contacts and control relay contacts are therefore available, although the number of actual input devices is limited. A finite number of actual outputs (such as Y2) are available, but their status can be read as many times as needed on other rungs of the ladder. Also, counters and timers are readily programmed on even the simplest PLC's.

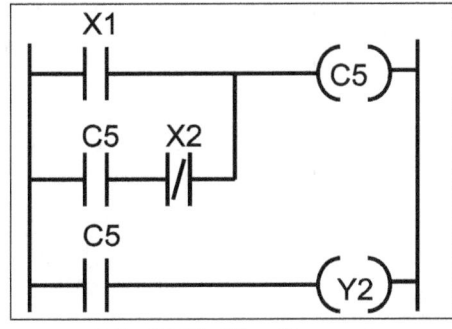

LC-: PLC ladder diagram.

In ME 360 we will use a combination of conventional hard-wired relay and PLC logic programming techniques. Rules for drawing ladder logic diagrams are summarized below:

- Ladder diagrams are drawn vertically with inputs on the left and outputs on the right.

- Each rung of the ladder has one (and only one) output.

- An individual output device can appear on the ladder diagram only once.

- An individual physical input device (limit switch, push-button, pressure switch, etc.) may be used as many times as necessary on the ladder diagram in both normally open and normally closed configurations, and is drawn using the schematic symbol of figure LC-.

- Internal contacts of the PLC are represented as conventional control relays and contacts.

- Control relay coils are outputs and can appear on the ladder diagram only once.

- Control relay contacts are inputs and may be used as many times as necessary on the ladder diagram in both normally open and normally closed configurations.

- Unlimited "OR"ing of ladder rungs is allowed, but any rung of the ladder diagram may be "OR"ed with a following rung at only one location.

Hard-wired logic systems are drawn in both horizontal and vertical configurations, but PLC diagrams are conventionally drawn vertically with single outputs in the rightmost column. If an output appeared at more than one location, then its status could be ambiguous. Physical input devices (limit switches, push-buttons, pressure switches, etc.) and relays have a limited number of normally open and normally closed contacts. Hard-wired logic systems can use only as many of these inputs or contacts as are physically available. PLC circuits allow unlimited use of their internal contacts which may also be connected to input devices. Contacts are normally shown for all inputs in a PLC diagram, but we will use the schematic symbols of figure LC- for clarity. Nested "OR" networks shown on the left in figure LC- are allowed in hard-wired relay logic systems, but are not in PLC's. Note that the first and second rungs are "OR"ed at two locations, as are the second and third rungs. The "OR" configuration shown on the right in figure LC- is allowed in a PLC system. Note that the order of the rungs on the right is irrelevant, the rungs are drawn in order of fewer components for clarity.

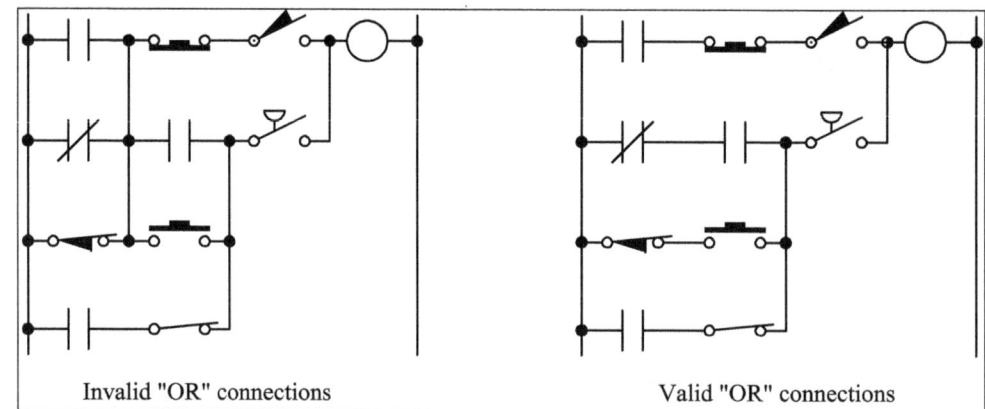

Invalid "OR" connections   Valid "OR" connections

LC-: Nested "OR" programming.

Logic Control Circuit Design

Designing new logic control circuits from "scratch" can be a daunting task. Oftentimes a designer can reuse logical blocks from previous successful designs. Unfortunately, this is not always possible, and can sometimes lead to unforseen interactions between various parts of the logic system. Some broad general guidelines (which are often violated!) for designing logic systems are given below:

1. Dedicate control relays for specific functions (such as starting the system, activating a solenoid, etc.) and use as many as are necessary. Control relays are essentially "free" once a programmable controller has been purchased, so don't be miserly.

2. Control relays almost always use a holding circuit, so design in terms of both a "turn ON" and a "turn OFF" rung with an "OR" connection between them. Note that some circuits will require multiple rungs for turning ON or OFF, which must be connected through the OR structure.

3. Normally open (N.O.) components are usually used to activate the "turn ON" rung.

4. Normally closed (N.C.) components are usually used to activate the "turn OFF" rung.

5. Be absolutely certain that any holding circuit formed will be actively turned off by your system. Do not depend on a power shutdown to release and holding circuits.

6. Provide safety interlocks either on the "turn ON" rung before the control relay or on the associated solenoid activation rung, depending on the type of interlock required.

7. Use the master control relay (MCR) concept for long-term effects, such as turning the entire system on, or starting a long sequence of actions.

Figure LC- below summarizes many of the design guidelines given above.

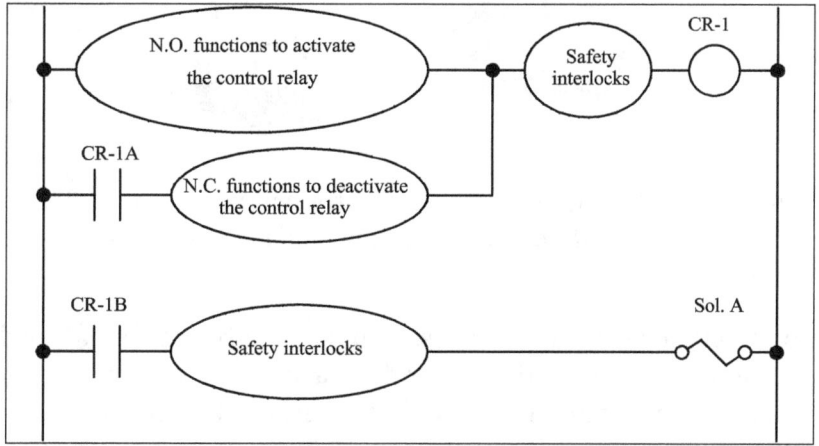

LC-. General logic design.

ON–OFF CONTROL

On-Off control is the simplest form of feedback control. An on-off controller simply drives the manipulated variable from fully closed to fully open depending on the position of the controlled variable relative to the setpoint. A common example of on-off control is the temperature control in a domestic heating system. When the temperature is below the thermostat setpoint the heating system is switched on and when the temperature is above the setpoint the heating switches off.

There is, however, a bit of subtlety applied in practical on-off systems. If the heating switches on and off the instant the measured temperature crossed the setpoint then the system would *chatter* - repeatedly switch on and off at very high frequency. If this happened the boiler wouldn't last very long! To avoid chattering, practical on-off controllers usually have a *deadband* around the set-point. When the measured value lies within this dead-band the controller does nothing - its only when the value moves outside that action is taken. The effect of this is to introduce continuous oscillation in the value of the controlled variable - the large the dead-band the higher the amplitude and lower the frequency.

Although on-off is a very cheap form of control it is rarely used in process control applications because of the oscillation it causes in the controlled and manipulated variables. In a connected process these oscillations would be propagated right through the system.

PROPORTIONAL CONTROL

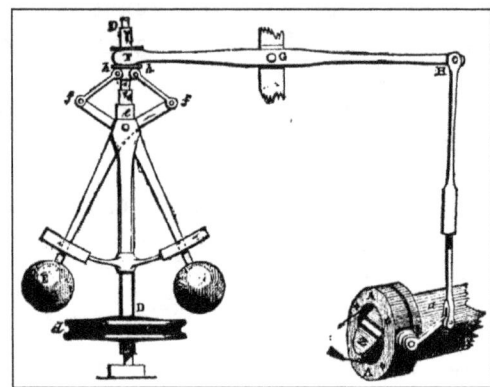

The fly-ball governor is an early classic example of proportional control. The balls rise as speed increases, which closes the valve, until a balance is achieved between demand and the proportional gain of the linkage and valve.

Proportional control, in engineering and process control, is a type of linear feedback control system in which a correction is applied to the controlled variable which is proportional to the difference between the desired value (setpoint, SP) and the measured value (process variable, PV). Two classic mechanical examples are the toilet bowl float proportioning valve and the fly-ball governor.

The proportional control concept is more complex than an on−off control system like a bi-metallic domestic thermostat, but simpler than a proportional−integral−derivative (PID) control system used in something like an automobile cruise control. On−off control will work where the overall system has a relatively long response time, but can result in instability if the system being controlled has a rapid response time. Proportional control overcomes this by modulating the output to the controlling device, such as a control valve at a level which avoids instability, but applies correction as fast as practicable by applying the optimum quantity of proportional gain.

A drawback of proportional control is that it cannot eliminate the residual SP − PV error in processes with compensation e.g. temperature control, as it requires an error to generate a proportional output. To overcome this the PI controller was devised, which uses a proportional term (P) to remove the gross error, and an integral term (I) to eliminate the residual offset error by integrating the error over time to produce an "I" component for the controller output.

Theory

In the proportional control algorithm, the controller output is proportional to the error signal, which is the difference between the setpoint and the process variable. In other words, the output of a proportional controller is the multiplication product of the error signal and the proportional gain.

This can be mathematically expressed as:

$$P_{out} = K_p\, e(t) + p0$$

where,

- $p0$: Controller output with zero error.

- P_{out} : Output of the proportional controller.

- K_p : Proportional gain.

- $e(t)$: Instantaneous process error at time t. $e(t) = SP - PV$

- SP : Set point.

- PV : Process variable.

Constraints: In a real plant, actuators have physical limitations that can be expressed as constraints on P_{out}. For example, P_{out} may be bounded between −1 and +1 if those are the maximum output limits.

Qualifications: It is preferable to express K_p as a unitless number. To do this, we can express $e(t)$ as a ratio with the span of the instrument. This span is in the same units as error (e.g. C degrees) so the ratio has no units.

Development of Control Block Diagrams

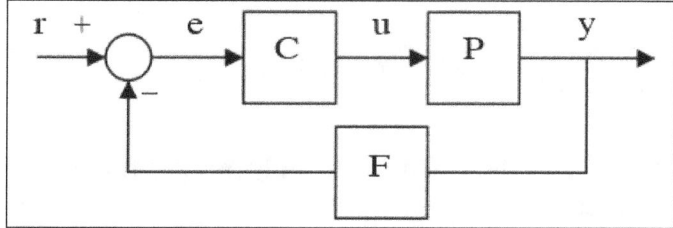

Simple feedback control loop2.

Proportional control dictates $g_c = k_c$. From the block diagram shown, assume that r, the setpoint, is the flowrate into a tank and e is *error*, which is the difference between setpoint and measured process output. g_p, is process transfer function; the input into the block is flow rate and output is tank level.

The output as a function of the setpoint, r, is known as the *closed-loop transfer function*. $g_{cl} = \dfrac{g_p g_c}{1 + g_p g_c}$, If the poles of g_{cl} are stable, then the closed-loop system is stable.

First-Order Process

For a first-order process, a general transfer function is $g_p = \dfrac{k_p}{\tau_p s + 1}$. Combining this with the

closed-loop transfer function above returns $g_{CL} = \dfrac{\dfrac{k_p k_c}{\tau_p s + 1}}{1 + \dfrac{k_p k_c}{\tau_p s + 1}}$. Simplifying this equation results in

$g_{CL} = \dfrac{k_{CL}}{\tau_{CL} s + 1}$ where $k_{CL} = \dfrac{k_p k_c}{1 + k_p k_c}$ and $\tau_{CL} = \dfrac{\tau_p}{1 + k_p k_c}$. For stability in this system, $\tau_{CL} > 0$; therefore,

τ_p must be a positive number, and $k_p k_c > -1$ (standard practice is to make sure that $k_p k_c > 0$).

Introducing a step change to the system gives the output response of $y(s) = g_{CL} \times \dfrac{\Delta R}{s}$.

Using the final-value theorem,

$$\lim_{t \to \infty} y(t) = \lim_{s \searrow 0}\left(s \times \frac{k_{CL}}{\tau_{CL} s + 1} \times \frac{\Delta R}{s} \right) = k_{CL} \times \Delta R = y(t)\big|_{t=\infty}$$

which shows that there will always be an offset in the system.

Integrating Process

For an integrating process, a general transfer function is $g_p = \dfrac{1}{s(s+1)}$, which, when combined with

the closed-loop transfer function, becomes $g_{CL} = \dfrac{k_c}{s(s+1) + k_c}$.

Introducing a step change to the system gives the output response of $y(s) = g_{CL} \times \dfrac{\Delta R}{s}$.

Using the final-value theorem,

$$\lim_{t \to \infty} y(t) = \lim_{s \searrow 0}\left(s \times \frac{k_c}{s(s+1) + k_c} \times \frac{\Delta R}{s} \right) = \Delta R = y(t)\big|_{t=\infty}$$

meaning there is no offset in this system. This is the only process that will not have any offset when using a proportional controller.

Offset Error

Proportional control cannot eliminate the offset error, which is the difference between the desired value and the actual value, SP – PV error, as it requires an error to generate an output. When a disturbance (deviation from existing state) occurs in the process value being controlled, any corrective control action, based purely on Proportional Control, will always leave out the error between the next steady state and the desired setpoint, and result in a residual error called the offset error. This error will increase as greater process demand is put on the system, or by increasing the set point.

Consider an object suspended by a spring as a simple proportional control. The spring will attempt to maintain the object in a certain location despite disturbances which may temporarily displace it. Hooke's law tells us that the spring applies a corrective force that is proportional to the object's displacement. While this will tend to hold the object in a particular location, the absolute resting location of the object will vary if its mass is changed. This difference in resting location is the offset error.

Imagine the same spring and object in a weightless environment. In this case, the spring will tend to hold the object in the same location regardless of its mass. There is no offset error in this case because the proportional action is not working against anything in the steady state.

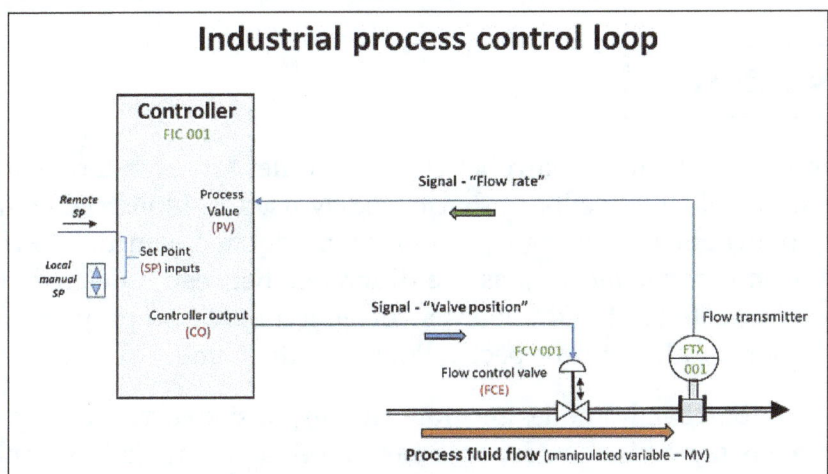

Flow control loop. If only a proportional controller, then there's always an offset between SP and PV.

Proportional Band

The proportional band is the band of controller output over which the final control element (a control valve, for instance) will move from one extreme to another. Mathematically, it can be expressed as:

$$PB = \frac{100}{K_p}$$

So if K_p, the proportional gain, is very high, the proportional band is very small, which means that the band of controller output over which the final control element will go from minimum to maximum (or vice versa) is very small. This is the case with on–off controllers, where K_p is very high and hence, for even a small error, the controller output is driven from one extreme to another.

Advantages

The clear advantage of proportional over on–off control can be demonstrated by car speed control. An analogy to on–off control is driving a car by applying either full power or no power and varying the duty cycle, to control speed. The power would be on until the target speed is reached, and then the power would be removed, so the car reduces speed. When the speed falls below the target, with a certain hysteresis, full power would again be applied. It can be seen that this would obviously result in poor control and large variations in speed. The more powerful the engine; the greater the instability, the heavier the car; the greater the stability. Stability may be expressed as correlating to the power-to-weight ratio of the vehicle.

In proportional control, the power output is always proportional to the (actual versus target speed) error. If the car is at target speed and the speed increases slightly due to a falling gradient, the power is reduced slightly, or in proportion to the change in error, so that the car reduces speed gradually and reaches the new target point with very little, if any, "overshoot", which is much smoother control than on–off control. In practice, PID controllers are used for this and the large number of control processes that require more response control than proportional alone.

PID CONTROL

A proportional–integral–derivative controller (PID controller. or three-term controller) is a control loop mechanism employing feedback that is widely used in industrial control systems and a variety of other applications requiring continuously modulated control. A PID controller continuously calculates an *error value* $e(t)$ as the difference between a desired setpoint (SP) and a measured process variable (PV) and applies a correction based on proportional, integral, and derivative terms (denoted P, I, and D respectively), hence the name.

In practical terms it automatically applies accurate and responsive correction to a control function. An everyday example is the cruise control on a car, where ascending a hill would lower speed if only constant engine power were applied. The controller's PID algorithm restores the measured speed to the desired speed with minimal delay and overshoot by increasing the power output of the engine.

The first theoretical analysis and practical application was in the field of automatic steering systems for ships, developed from the early 1920s onwards. It was then used for automatic process control in the manufacturing industry, where it was widely implemented in pneumatic, and then electronic, controllers. Today the PID concept is used universally in applications requiring accurate and optimised automatic control.

Fundamental Operation

The distinguishing feature of the PID controller is the ability to use the three *control terms* of proportional, integral and derivative influence on the controller output to apply accurate and optimal control. The block diagram on the right shows the principles of how these terms are generated and applied. It shows a PID controller, which continuously calculates an *error value* $e(t)$ as the difference between a desired setpoint $\text{SP} = r(t)$ and a measured process variable $\text{PV} = y(t)$, and applies a correction based on proportional, integral, and derivative terms. The controller attempts to minimize the error over time by adjustment of a *control variable* $u(t)$, such as the opening of a control valve, to a new value determined by a weighted sum of the control terms.

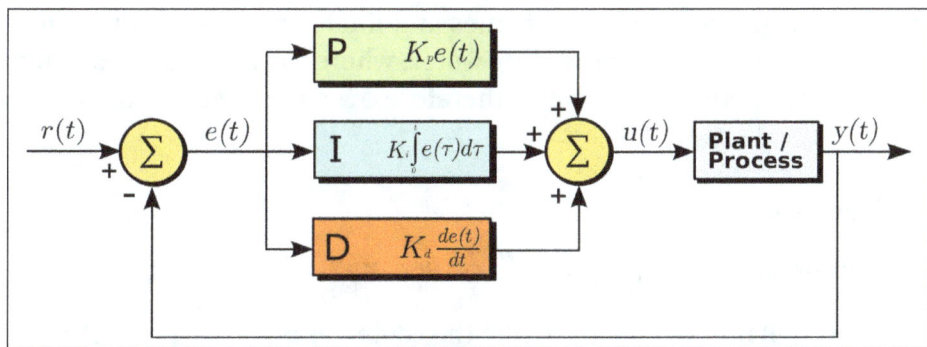

A block diagram of a PID controller in a feedback loop. $r(t)$ is the desired process value or setpoint (SP), and $y(t)$ is the measured process value (PV).

In this model:

- Term P is proportional to the current value of the SP − PV error e(t). For example, if the error is large and positive, the control output will be proportionately large and positive, taking into account the gain factor "K". Using proportional control alone will result in an error between the setpoint and the actual process value, because it requires an error to generate the proportional response. If there is no error, there is no corrective response.

- Term I accounts for past values of the SP − PV error and integrates them over time to produce the I term. For example, if there is a residual SP − PV error after the application of proportional control, the integral term seeks to eliminate the residual error by adding a control effect due to the historic cumulative value of the error. When the error is eliminated, the integral term will cease to grow. This will result in the proportional effect diminishing as the error decreases, but this is compensated for by the growing integral effect.

- Term D is a best estimate of the future trend of the SP − PV error, based on its current rate of change. It is sometimes called "anticipatory control", as it is effectively seeking to reduce the effect of the SP − PV error by exerting a control influence generated by the rate of error change. The more rapid the change, the greater the controlling or dampening effect.

Tuning − The balance of these effects is achieved by loop tuning to produce the optimal control function. The tuning constants are shown below as "K" and must be derived for each control application, as they depend on the response characteristics of the complete loop external to the controller. These are dependent on the behaviour of the measuring sensor, the final control element (such as a control valve), any control signal delays and the process itself. Approximate values of constants can usually be initially entered knowing the type of application, but they are normally refined, or tuned, by "bumping" the process in practice by introducing a setpoint change and observing the system response.

Control action − The mathematical model and practical loop above both use a "direct" control action for all the terms, which means an increasing positive error results in an increasing positive control output for the summed terms to apply correction. However, the output is called "reverse" acting if it is necessary to apply negative corrective action. For instance, if the valve in the flow loop was 100−0% valve opening for 0−100% control output − meaning that the controller action has to be reversed. Some process control schemes and final control elements require this reverse action. An example would be a valve for cooling water, where the fail-safe mode, in the case of loss of signal, would be 100% opening of the valve; therefore 0% controller output needs to cause 100% valve opening.

Mathematical Form

The overall control function $u(t) = K_{\mathrm{p}} e(t) + K_{\mathrm{i}} \int_0^t e(t') dt' + K_{\mathrm{d}} \dfrac{de(t)}{dt}$,

where K_{p}, K_{i}, and K_{d}, all non-negative, denote the coefficients for the proportional, integral, and derivative terms respectively (sometimes denoted P, I, and D).

In the *standard form* of the equation, K_{i} and K_{d} are respectively replaced by $K_{\mathrm{p}} / T_{\mathrm{i}}$ and $K_{\mathrm{p}} T_{\mathrm{d}}$; the advantage of this being that T_{i} and T_{d} have some understandable physical meaning, as they represent the integration time and the derivative time respectively.

$$u(t) = K_{\mathrm{p}} \left(e(t) + \frac{1}{T_{\mathrm{i}}} \int_0^t e(t') dt' + T_{\mathrm{d}} \frac{de(t)}{dt} \right),$$

Selective use of Control Terms

Although a PID controller has three control terms, some applications use only one or two terms to provide the appropriate control. This is achieved by setting the unused parameters to zero and is called a PI, PD, P or I controller in the absence of the other control actions. PI controllers are fairly common, since derivative action is sensitive to measurement noise, whereas the absence of an integral term may prevent the system from reaching its target value.

Applicability

The use of the PID algorithm does not guarantee optimal control of the system or its control stability. Situations may occur where there are excessive delays: the measurement of the process value is delayed, or the control action does not apply quickly enough. In these cases lead–lag compensation is required to be effective. The response of the controller can be described in terms of its responsiveness to an error, the degree to which the system overshoots a setpoint, and the degree of any system oscillation. But the PID controller is broadly applicable, since it relies only on the response of the measured process variable, not on knowledge or a model of the underlying process.

Control Loop Example

Consider a robotic arm that can be moved and positioned by a control loop. An electric motor may lift or lower the arm, depending on forward or reverse power applied, but power cannot be a simple function of position because of the inertial mass of the arm, forces due to gravity, external forces on the arm such as a load to lift or work to be done on an external object.

- The sensed position is the process variable (PV).

- The desired position is called the setpoint (SP).

- The difference between the PV and SP is the error (e), which quantifies whether the arm is too low or too high and by how much.

- The input to the process (the electric current in the motor) is the output from the PID controller. It is called either the manipulated variable (MV) or the control variable (CV).

By measuring the position (PV), and subtracting it from the setpoint (SP), the error (e) is found, and from it the controller calculates how much electric current to supply to the motor (MV).

Proportional

The obvious method is proportional control: the motor current is set in proportion to the existing error. However, this method fails if, for instance, the arm has to lift different weights: a greater weight needs a greater force applied for a same error on the down side, but a smaller force if the error is on the upside. That's where the integral and derivative terms play their part.

Integral

An integral term increases action in relation not only to the error but also the time for which it has persisted. So, if applied force is not enough to bring the error to zero, this force will be increased as time passes. A pure "I" controller could bring the error to zero, but it would be both slow reacting at the start (because action would be small at the beginning, needing time to get significant) and brutal (the action increases as long as the error is positive, even if the error has started to approach zero).

Derivative

A derivative term does not consider the error (meaning it cannot bring it to zero: a pure D controller cannot bring the system to its setpoint), but the rate of change of error, trying to bring this rate to zero. It aims at flattening the error trajectory into a horizontal line, damping the force applied, and so reduces overshoot (error on the other side because too great applied force). Applying too much impetus when the error is small and decreasing will lead to overshoot. After overshooting, if the controller were to apply a large correction in the opposite direction and repeatedly overshoot the desired position, the output would oscillate around the setpoint in either a constant, growing, or decaying sinusoid. If the amplitude of the oscillations increases with time, the system is unstable. If they decrease, the system is stable. If the oscillations remain at a constant magnitude, the system is marginally stable.

Control Damping

In the interest of achieving a controlled arrival at the desired position (SP) in a timely and accurate way, the controlled system needs to be critically damped. A well-tuned position control system will also apply the necessary currents to the controlled motor so that the arm pushes and pulls as necessary to resist external forces trying to move it away from the required position. The setpoint itself may be generated by an external system, such as a PLC or other computer system, so that it continuously varies depending on the work that the robotic arm is expected to do. A well-tuned PID control system will enable the arm to meet these changing requirements to the best of its capabilities.

Response to Disturbances

If a controller starts from a stable state with zero error (PV = SP), then further changes by the controller will be in response to changes in other measured or unmeasured inputs to the process that affect the process, and hence the PV. Variables that affect the process other than the MV are known as disturbances. Generally, controllers are used to reject disturbances and to implement setpoint changes. A change in load on the arm constitutes a disturbance to the robot arm control process.

Applications

In theory, a controller can be used to control any process which has a measurable output (PV), a known ideal value for that output (SP) and an input to the process (MV) that will affect the relevant PV. Controllers are used in industry to regulate temperature, pressure, force, feed rate, flow rate, chemical composition (component concentrations), weight, position, speed, and practically every other variable for which a measurement exists.

PID Controller Theory

The PID control scheme is named after its three correcting terms, whose sum constitutes the manipulated variable (MV). The proportional, integral, and derivative terms are summed to calculate the output of the PID controller. Defining $u(t)$ as the controller output, the final form of the PID algorithm is:

$$u(t) = \text{MV}(t) = K_p e(t) + K_i \int_0^t e(\tau)d\tau + K_d \frac{de(t)}{dt},$$

where,

- K_p is the proportional gain, a tuning parameter.

- K_i is the integral gain, a tuning parameter.

- K_d is the derivative gain, a tuning parameter.

- $e(t) = \text{SP} - \text{PV}(t)$ is the error (SP is the setpoint, and PV(t) is the process variable).

- t is the time or instantaneous time (the present).

- τ is the variable of integration (takes on values from time 0 to the present t).

Equivalently, the transfer function in the Laplace domain of the PID controller is:

$$L(s) = K_p + K_i / s + K_d s,$$

where, s is the complex frequency.

Proportional Term

The proportional term produces an output value that is proportional to the current error value. The proportional response can be adjusted by multiplying the error by a constant K_p, called the proportional gain constant.

The proportional term is given by,

$$P_{out} = K_p e(t).$$

A high proportional gain results in a large change in the output for a given change in the error. If the proportional gain is too high, the system can become unstable (see the section on loop tuning). In contrast, a small gain results in a small output response to a large input error, and a less responsive or less sensitive controller. If the proportional gain is too low, the control action may be too small when responding to system disturbances. Tuning theory and industrial practice indicate that the proportional term should contribute the bulk of the output change.

Steady-state Error

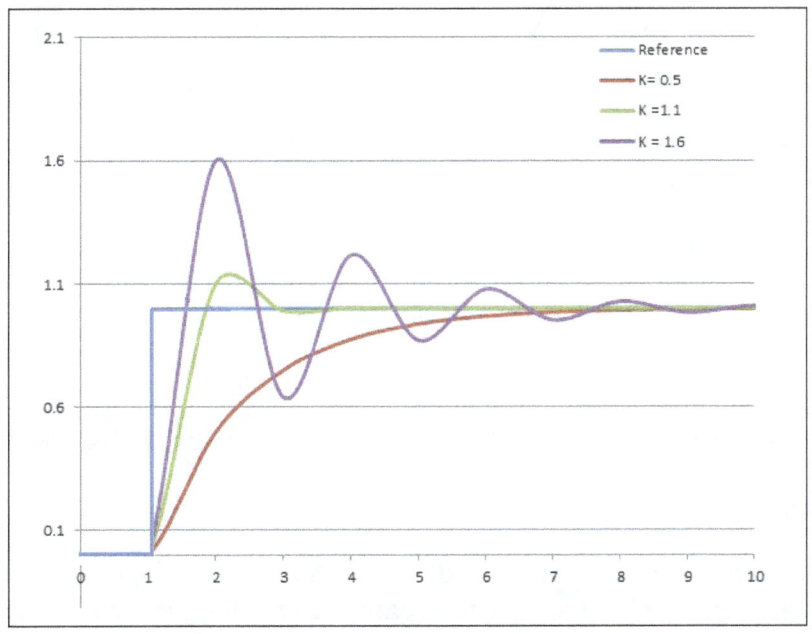

Response of PV to step change of SP vs time, for three values of K_p (K_i and K_d held constant).

The steady-state error is the difference between the desired final output and the actual one. Because a non-zero error is required to drive it, a proportional controller generally operates with a steady-state error. Steady-state error (SSE) is proportional to the process gain and inversely proportional to proportional gain. SSE may be mitigated by adding a compensating bias term to the setpoint AND output, or corrected dynamically by adding an integral term.

Integral Term

The contribution from the integral term is proportional to both the magnitude of the error and the duration of the error. The integral in a PID controller is the sum of the instantaneous error over time and gives the accumulated offset that should have been corrected previously. The accumulated error is then multiplied by the integral gain (K_i) and added to the controller output.

The integral term is given by

$$I_{out} = K_i \int_0^t e(\tau)d\tau.$$

The integral term accelerates the movement of the process towards setpoint and eliminates the residual steady-state error that occurs with a pure proportional controller. However, since the integral term responds to accumulated errors from the past, it can cause the present value to overshoot the setpoint value.

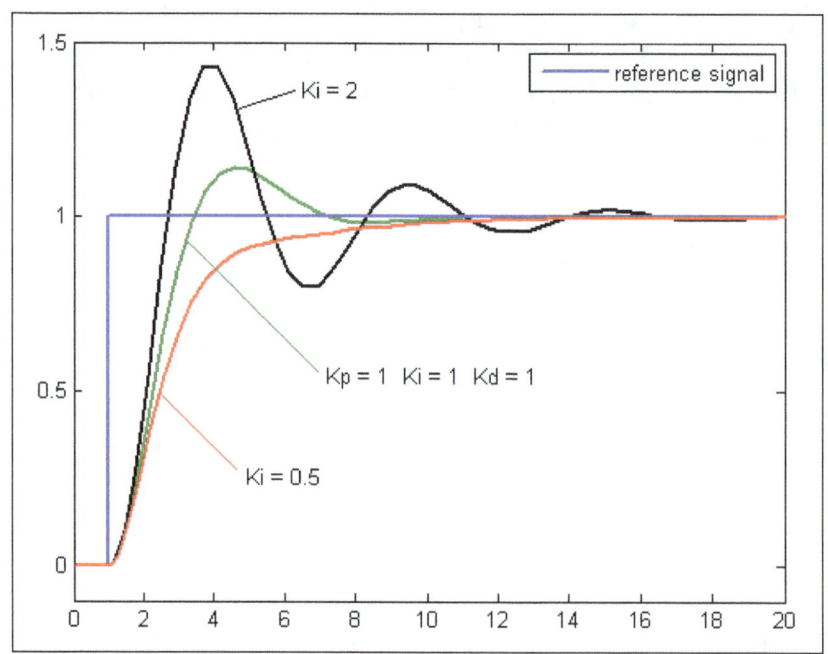

Response of PV to step change of SP vs time, for three values of K_i (K_p and K_d held constant).

Derivative Term

The derivative of the process error is calculated by determining the slope of the error over time and multiplying this rate of change by the derivative gain K_d. The magnitude of the contribution of the derivative term to the overall control action is termed the derivative gain, K_d.

The derivative term is given by,

$$D_{out} = K_d \frac{de(t)}{dt}.$$

Derivative action predicts system behavior and thus improves settling time and stability of the system. An ideal derivative is not causal, so that implementations of PID controllers include an additional low-pass filtering for the derivative term to limit the high-frequency gain and noise. Derivative action is seldom used in practice though – by one estimate in only 25% of deployed controllers – because of its variable impact on system stability in real-world applications.

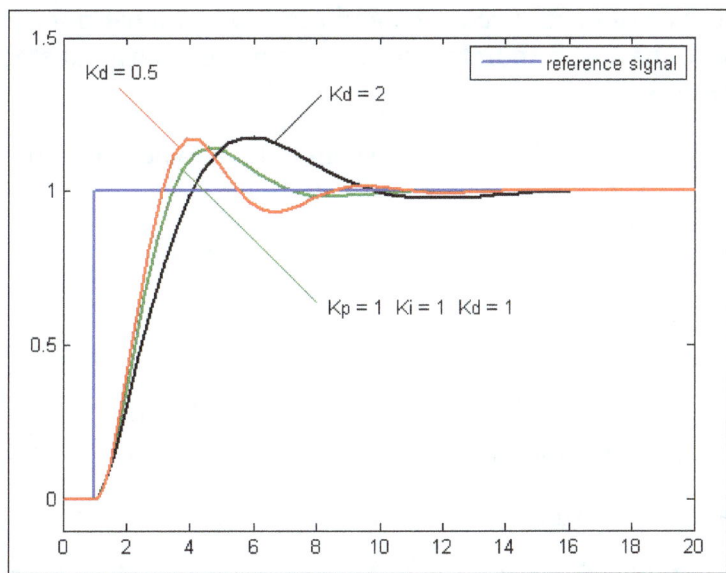

Response of PV to step change of SP vs time, for three values of K_d (K_p and K_i held constant).

Loop Tuning

Tuning a control loop is the adjustment of its control parameters (proportional band/gain, integral gain/reset, derivative gain/rate) to the optimum values for the desired control response. Stability (no unbounded oscillation) is a basic requirement, but beyond that, different systems have different behavior, different applications have different requirements, and requirements may conflict with one another.

PID tuning is a difficult problem, even though there are only three parameters and in principle is simple to describe, because it must satisfy complex criteria within the limitations of PID control. There are accordingly various methods for loop tuning, and more sophisticated techniques are the subject of patents.

Designing and tuning a PID controller appears to be conceptually intuitive, but can be hard in practice, if multiple (and often conflicting) objectives such as short transient and high stability are to be achieved. PID controllers often provide acceptable control using default tunings, but performance can generally be improved by careful tuning, and performance may be unacceptable with poor tuning. Usually, initial designs need to be adjusted repeatedly through computer simulations until the closed-loop system performs or compromises as desired.

Some processes have a degree of nonlinearity and so parameters that work well at full-load conditions don't work when the process is starting up from no-load; this can be corrected by gain scheduling (using different parameters in different operating regions).

Stability

If the PID controller parameters (the gains of the proportional, integral and derivative terms) are chosen incorrectly, the controlled process input can be unstable, i.e., its output diverges, with or without oscillation, and is limited only by saturation or mechanical breakage. Instability is caused by *excess* gain, particularly in the presence of significant lag.

Generally, stabilization of response is required and the process must not oscillate for any combination of process conditions and setpoints, though sometimes marginal stability (bounded oscillation) is acceptable or desired.

Mathematically, the origins of instability can be seen in the Laplace domain.

The total loop transfer function is:

$$H(s) = \frac{K(s)G(s)}{1 + K(s)G(s)}$$

where,

- $K(s)$ is the PID transfer function.

- $G(s)$ is the plant transfer function.

The system is called unstable where the closed loop transfer function diverges for some s. This happens for situations where $K(s)G(s) = -1$. Typically, this happens when $|K(s)G(s)| = 1$ with a 180 degree phase shift. Stability is guaranteed when $K(s)G(s) < 1$ for frequencies that suffer high phase shifts. A more general formalism of this effect is known as the Nyquist stability criterion.

Optimal Behavior

The optimal behavior on a process change or setpoint change varies depending on the application.

Two basic requirements are *regulation* (disturbance rejection – staying at a given setpoint) and *command tracking* (implementing setpoint changes) – these refer to how well the controlled variable tracks the desired value. Specific criteria for command tracking include rise time and settling time. Some processes must not allow an overshoot of the process variable beyond the setpoint if, for example, this would be unsafe. Other processes must minimize the energy expended in reaching a new setpoint.

Tuning Methods

There are several methods for tuning a PID loop. The most effective methods generally involve the development of some form of process model, then choosing P, I, and D based on the dynamic model parameters. Manual tuning methods can be relatively time consuming, particularly for systems with long loop times.

The choice of method will depend largely on whether or not the loop can be taken offline for tuning, and on the response time of the system. If the system can be taken offline, the best tuning method often involves subjecting the system to a step change in input, measuring the output as a function of time, and using this response to determine the control parameters.

Choosing a tuning method		
Method	Advantages	Disadvantages
Manual tuning	No math required; online.	Requires experienced personnel.

Ziegler–Nichols	Proven method; online.	Process upset, some trial-and-error, very aggressive tuning.
Tyreus Luyben	Proven method; online.	Process upset, some trial-and-error, very aggressive tuning.
Software tools	Consistent tuning; online or offline - can employ computer-automated control system design (CAutoD) techniques; may include valve and sensor analysis; allows simulation before downloading; can support non-steady-state (NSS) tuning.	Some cost or training involved.
Cohen–Coon	Good process models.	Some math; offline; only good for first-order processes.
Åström-Hägglund	Can be used for auto tuning; amplitude is minimum so this method has lowest process upset	The process itself is inherently oscillatory.

Manual Tuning

If the system must remain online, one tuning method is to first set K_i and K_d values to zero. Increase the K_p until the output of the loop oscillates, then the K_p should be set to approximately half of that value for a "quarter amplitude decay" type response. Then increase K_i until any offset is corrected in sufficient time for the process. However, too much K_i will cause instability. Finally, increase K_d, if required, until the loop is acceptably quick to reach its reference after a load disturbance. However, too much K_d will cause excessive response and overshoot. A fast PID loop tuning usually overshoots slightly to reach the setpoint more quickly; however, some systems cannot accept overshoot, in which case an overdamped closed-loop system is required, which will require a K_p setting significantly less than half that of the K_p setting that was causing oscillation.

Effects of *increasing* a parameter independently				
Parameter	Rise time	Overshoot	Settling time	Steady-state error	Stability
K_p	Decrease	Increase	Small change	Decrease	Degrade
K_i	Decrease	Increase	Increase	Eliminate	Degrade
K_d	Minor change	Decrease	Decrease	No effect in theory	Improve if K_d small

Ziegler–Nichols Method

Another heuristic tuning method is formally known as the Ziegler–Nichols method, introduced by John G. Ziegler and Nathaniel B. Nichols in the 1940s. As in the method above, the K_i and K_d gains are first set to zero. The proportional gain is increased until it reaches the ultimate gain, K_u, at which the output of the loop starts to oscillate. K_u and the oscillation period T_u are used to set the gains as follows:

Ziegler–Nichols method			
Control Type	K_p	K_i	K_d
P	$0.50K_u$	—	—

PI	$0.45K_u$	$0.54K_u/T_u$	—
PID	$0.60K_u$	$1.2K_u/T_u$	$3K_uT_u/40$

These gains apply to the ideal, parallel form of the PID controller. When applied to the standard PID form, only the integral and derivative time parameters T_i and T_d are dependent on the oscillation period T_u.

Cohen–Coon Parameters

This method was developed in 1953 and is based on a first-order + time delay model. Similar to the Ziegler–Nichols method, a set of tuning parameters were developed to yield a closed-loop response with a decay ratio of 1/4. Arguably the biggest problem with these parameters is that a small change in the process parameters could potentially cause a closed-loop system to become unstable.

Relay (Åström–Hägglund) Method

Published in 1984 by Karl Johan Åström and Tore Hägglund, the relay method temporarily operates the process using bang-bang control and measures the resultant oscillations. The output is switched (as if by a relay, hence the name) between two values of the control variable. The values must be chosen so the process will cross the setpoint, but need not be 0% and 100%; by choosing suitable values, dangerous oscillations can be avoided.

As long as the process variable is below the setpoint, the control output is set to the higher value. As soon as it rises above the setpoint, the control output is set to the lower value. Ideally, the output waveform is nearly square, spending equal time above and below the setpoint. The period and amplitude of the resultant oscillations are measured, and used to compute the ultimate gain and period, which are then fed into the Ziegler–Nichols method.

Specifically, the ultimate period T_u is assumed to be equal to the observed period, and the ultimate gain is computed as $K_u = 4b/\pi a$, where a is the amplitude of the process variable oscillation, and b is the amplitude of the control output change which caused it.

There are numerous variants on the relay method.

First Order + Dead Time Model

The transfer function for a first order process, with dead time, is:

$$y(s) = \frac{k_p e^{-\theta s}}{\tau_p s + 1} * u(s)$$

where,

- k_p is the process gain.

- τ_p is the time constant.

- θ is the dead time.

- u(s) is a step change input.

Converting this transfer function to the time domain results in:

$$y(t) = k_p \Delta u (1 - e^{\frac{-t-\theta}{\tau_p}})$$

using the same parameters found above.

It is important when using this method to apply a large enough step change input that the output can be measured; however, too large of a step change can affect the process stability. Additionally, a larger step change will ensure that the output is not changing due to a disturbance (for best results, try to minimize disturbances when performing the step test).

One way to determine the parameters for the first order process is using the 63.2% method. In this method, the process gain (k_p) is equal to the change in output divided by the change in input. The dead time (θ) is the amount of time between when the step change occurred and when the output first changed. The time constant (τ_p) is the amount of time it takes for the output to reach 63.2% of the new steady state value after the step change. One downside to using this method is that the time to reach a new steady state value can take a while if the process has a large time constants.

PID Tuning Software

Most modern industrial facilities no longer tune loops using the manual calculation methods shown above. Instead, PID tuning and loop optimization software are used to ensure consistent results. These software packages will gather the data, develop process models, and suggest optimal tuning. Some software packages can even develop tuning by gathering data from reference changes.

Mathematical PID loop tuning induces an impulse in the system, and then uses the controlled system's frequency response to design the PID loop values. In loops with response times of several minutes, mathematical loop tuning is recommended, because trial and error can take days just to find a stable set of loop values. Optimal values are harder to find. Some digital loop controllers offer a self-tuning feature in which very small setpoint changes are sent to the process, allowing the controller itself to calculate optimal tuning values.

Another approach calculates initial values via the Ziegler–Nichols method, and uses a numerical optimization technique to find better PID coefficients.

Other formulas are available to tune the loop according to different performance criteria. Many patented formulas are now embedded within PID tuning software and hardware modules.

Advances in automated PID loop tuning software also deliver algorithms for tuning PID Loops in a dynamic or non-steady state (NSS) scenario. The software will model the dynamics of a process, through a disturbance, and calculate PID control parameters in response.

Limitations of PID control

While PID controllers are applicable to many control problems, and often perform satisfactorily without any improvements or only coarse tuning, they can perform poorly in some applications, and do not in general provide *optimal* control. The fundamental difficulty with PID control is that it is a feedback control system, with *constant* parameters, and no direct knowledge of the process, and thus overall performance is reactive and a compromise. While PID control is the best controller in an observer without a model of the process, better performance can be obtained by overtly modeling the actor of the process without resorting to an observer.

PID controllers, when used alone, can give poor performance when the PID loop gains must be reduced so that the control system does not overshoot, oscillate or hunt about the control setpoint value. They also have difficulties in the presence of non-linearities, may trade-off regulation versus response time, do not react to changing process behavior (say, the process changes after it has warmed up), and have lag in responding to large disturbances.

The most significant improvement is to incorporate feed-forward control with knowledge about the system, and using the PID only to control error. Alternatively, PIDs can be modified in more minor ways, such as by changing the parameters (either gain scheduling in different use cases or adaptively modifying them based on performance), improving measurement (higher sampling rate, precision, and accuracy, and low-pass filtering if necessary), or cascading multiple PID controllers.

Linearity

Another problem faced with PID controllers is that they are linear, and in particular symmetric. Thus, performance of PID controllers in non-linear systems (such as HVAC systems) is variable. For example, in temperature control, a common use case is active heating (via a heating element) but passive cooling (heating off, but no cooling), so overshoot can only be corrected slowly – it cannot be forced downward. In this case the PID should be tuned to be overdamped, to prevent or reduce overshoot, though this reduces performance (it increases settling time).

Noise In Derivative

A problem with the derivative term is that it amplifies higher frequency measurement or process noise that can cause large amounts of change in the output. It is often helpful to filter the measurements with a low-pass filter in order to remove higher-frequency noise components. As low-pass filtering and derivative control can cancel each other out, the amount of filtering is limited. Therefore, low noise instrumentation can be important. A nonlinear median filter may be used, which improves the filtering efficiency and practical performance. In some cases, the differential band can be turned off with little loss of control. This is equivalent to using the PID controller as a PI controller.

Modifications to the PID Algorithm

The basic PID algorithm presents some challenges in control applications that have been addressed by minor modifications to the PID form.

Integral Windup

One common problem resulting from the ideal PID implementations is integral windup. Following a large change in setpoint the integral term can accumulate an error larger than the maximal value for the regulation variable (windup), thus the system overshoots and continues to increase until this accumulated error is unwound. This problem can be addressed by:

- Disabling the integration until the PV has entered the controllable region.
- Preventing the integral term from accumulating above or below pre-determined bounds.
- Back-calculating the integral term to constrain the regulator output within feasible bounds.

Overshooting from Known Disturbances

For example, a PID loop is used to control the temperature of an electric resistance furnace where the system has stabilized. Now when the door is opened and something cold is put into the furnace the temperature drops below the setpoint. The integral function of the controller tends to compensate for error by introducing another error in the positive direction. This overshoot can be avoided by freezing of the integral function after the opening of the door for the time the control loop typically needs to reheat the furnace.

PI Controller

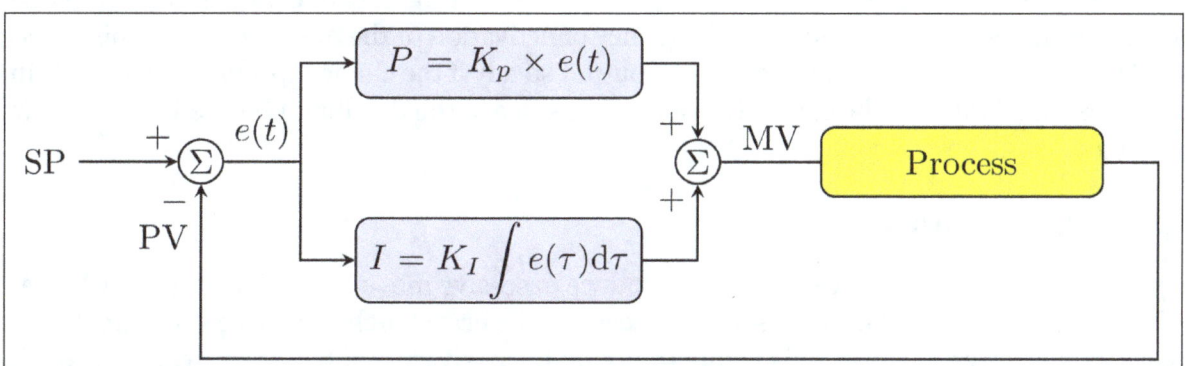

Basic block of a PI controller.

A PI controller (proportional-integral controller) is a special case of the PID controller in which the derivative (D) of the error is not used.

The controller output is given by,

$$K_p \Delta + K_I \int \Delta dt$$

where, Δ is the error or deviation of actual measured value (*PV*) from the setpoint (*SP*).

$$\Delta = SP - PV.$$

A PI controller can be modelled easily in software such as Simulink or Xcos using a "flow chart" box involving Laplace operators:

$$C = \frac{G(1 + \tau s)}{\tau s}$$

where,

$$G = K_P = \text{proportional gain}$$
$$\frac{G}{\tau} = K_I = \text{integral gain}$$

Setting a value for G is often a trade off between decreasing overshoot and increasing settling time.

The lack of derivative action may make the system more steady in the steady state in the case of noisy data. This is because derivative action is more sensitive to higher-frequency terms in the inputs.

Without derivative action, a PI-controlled system is less responsive to real (non-noise) and relatively fast alterations in state and so the system will be slower to reach setpoint and slower to respond to perturbations than a well-tuned PID system may be.

Deadband

Many PID loops control a mechanical device (for example, a valve). Mechanical maintenance can be a major cost and wear leads to control degradation in the form of either stiction or backlash in the mechanical response to an input signal. The rate of mechanical wear is mainly a function of how often a device is activated to make a change. Where wear is a significant concern, the PID loop may have an output deadband to reduce the frequency of activation of the output (valve). This is accomplished by modifying the controller to hold its output steady if the change would be small (within the defined deadband range). The calculated output must leave the deadband before the actual output will change.

Setpoint Step Change

The proportional and derivative terms can produce excessive movement in the output when a system is subjected to an instantaneous step increase in the error, such as a large setpoint change. In the case of the derivative term, this is due to taking the derivative of the error, which is very large in the case of an instantaneous step change. As a result, some PID algorithms incorporate some of the following modifications:

- Setpoint ramping: In this modification, the setpoint is gradually moved from its old value to a newly specified value using a linear or first order differential ramp function. This avoids the discontinuity present in a simple step change.

- Derivative of the process variable: In this case the PID controller measures the derivative of the measured process variable (PV), rather than the derivative of the error. This quantity is always continuous (i.e., never has a step change as a result of changed setpoint). This modification is a simple case of setpoint weighting.

- Setpoint weighting: Setpoint weighting adds adjustable factors (usually between 0 and 1) to the setpoint in the error in the proportional and derivative element of the controller. The error in the integral term must be the true control error to avoid steady-state control errors. These two extra parameters do not affect the response to load disturbances and measurement noise and can be tuned to improve the controller's setpoint response.

Feed-Forward

The control system performance can be improved by combining the feedback (or closed-loop) control of a PID controller with feed-forward (or open-loop) control. Knowledge about the system (such as the desired acceleration and inertia) can be fed forward and combined with the PID output to improve the overall system performance. The feed-forward value alone can often provide the major portion of the controller output. The PID controller primarily has to compensate whatever difference or *error* remains between the setpoint (SP) and the system response to the open loop control. Since the feed-forward output is not affected by the process feedback, it can never cause the control system to oscillate, thus improving the system response without affecting stability. Feed forward can be based on the setpoint and on extra measured disturbances. Setpoint weighting is a simple form of feed forward.

For example, in most motion control systems, in order to accelerate a mechanical load under control, more force is required from the actuator. If a velocity loop PID controller is being used to control the speed of the load and command the force being applied by the actuator, then it is beneficial to take the desired instantaneous acceleration, scale that value appropriately and add it to the output of the PID velocity loop controller. This means that whenever the load is being accelerated or decelerated, a proportional amount of force is commanded from the actuator regardless of the feedback value. The PID loop in this situation uses the feedback information to change the combined output to reduce the remaining difference between the process setpoint and the feedback value. Working together, the combined open-loop feed-forward controller and closed-loop PID controller can provide a more responsive control system.

Bumpless Operation

PID controllers are often implemented with a "bumpless" initialization feature that recalculates the integral accumulator term to maintain a consistent process output through parameter changes. A partial implementation is to store the integral of the integral gain times the error rather than storing the integral of the error and postmultiplying by the integral gain, which prevents discontinuous output when the I gain is changed, but not the P or D gains.

Other Improvements

In addition to feed-forward, PID controllers are often enhanced through methods such as PID gain scheduling (changing parameters in different operating conditions), fuzzy logic, or computational verb logic. Further practical application issues can arise from instrumentation connected to the controller. A high enough sampling rate, measurement precision, and measurement accuracy are required to achieve adequate control performance. Another new method for improvement of PID controller is to increase the degree of freedom by using fractional order. The order of the integrator and differentiator add increased flexibility to the controller.

Cascade Control

One distinctive advantage of PID controllers is that two PID controllers can be used together to yield better dynamic performance. This is called cascaded PID control. In cascade control there are two PIDs arranged with one PID controlling the setpoint of another. A PID controller acts as outer loop controller, which controls the primary physical parameter, such as fluid level or velocity. The other controller acts as inner loop controller, which reads the output of outer loop controller as setpoint, usually controlling a more rapid changing parameter, flowrate or acceleration. It can be mathematically proven that the working frequency of the controller is increased and the time constant of the object is reduced by using cascaded PID controllers.

For example, a temperature-controlled circulating bath has two PID controllers in cascade, each with its own thermocouple temperature sensor. The outer controller controls the temperature of the water using a thermocouple located far from the heater, where it accurately reads the temperature of the bulk of the water. The error term of this PID controller is the difference between the desired bath temperature and measured temperature. Instead of controlling the heater directly, the outer PID controller sets a heater temperature goal for the inner PID controller. The inner PID controller controls the temperature of the heater using a thermocouple attached to the heater. The inner controller's error term is the difference between this heater temperature setpoint and the measured temperature of the heater. Its output controls the actual heater to stay near this setpoint.

The proportional, integral, and differential terms of the two controllers will be very different. The outer PID controller has a long time constant – all the water in the tank needs to heat up or cool down. The inner loop responds much more quickly. Each controller can be tuned to match the physics of the system *it* controls – heat transfer and thermal mass of the whole tank or of just the heater – giving better total response.

Alternative Nomenclature and PID Forms

Standard vs. Parallel (ideal) PID Form

The form of the PID controller most often encountered in industry, and the one most relevant to tuning algorithms is the *standard form*. In this form the K_p gain is applied to the I_{out}, and D_{out} terms, yielding:

$$u(t) = K_p \left(e(t) + \frac{1}{T_i} \int_0^t e(\tau)d\tau + T_d \frac{d}{dt} e(t) \right)$$

where,

- T_i is the integral time.

- T_d is the derivative time.

In this standard form, the parameters have a clear physical meaning. In particular, the inner summation produces a new single error value which is compensated for future and past errors. The proportional error term is the current error. The derivative components term attempts to predict the error value at T_d seconds (or samples) in the future, assuming that the loop control remains unchanged. The integral component adjusts the error value to compensate for the sum of all past errors, with the intention of completely eliminating them in T_i seconds (or samples). The resulting compensated single error value is then scaled by the single gain K_p to compute the control variable.

In the parallel form, shown in the controller theory section

$$u(t) = K_p e(t) + K_i \int_0^t e(\tau)d\tau + K_d \frac{d}{dt}e(t)$$

the gain parameters are related to the parameters of the standard form through $K_i = K_p / T_i$ and $K_d = K_p T_d$. This parallel form, where the parameters are treated as simple gains, is the most general and flexible form. However, it is also the form where the parameters have the least physical interpretation and is generally reserved for theoretical treatment of the PID controller. The standard form, despite being slightly more complex mathematically, is more common in industry.

Reciprocal Gain

In many cases, the manipulated variable output by the PID controller is a dimensionless fraction between 0 and 100% of some maximum possible value, and the translation into real units (such as pumping rate or watts of heater power) is outside the PID controller. The process variable, however, is in dimensioned units such as temperature. It is common in this case to express the gain K_p not as "output per degree", but rather in the reciprocal form of a *proportional band* $100 / K_p$, which is "degrees per full output": the range over which the output changes from 0 to 1 (0% to 100%). Beyond this range, the output is saturated, full-off or full-on. The narrower this band, the higher the proportional gain.

Basing Derivative Action on PV

In most commercial control systems, derivative action is based on process variable rather than error. That is, a change in the setpoint does not affect the derivative action. This is because the digitized version of the algorithm produces a large unwanted spike when the setpoint is changed. If the setpoint is constant then changes in the PV will be the same as changes in error. Therefore, this modification makes no difference to the way the controller responds to process disturbances.

Basing Proportional Action on PV

Most commercial control systems offer the *option* of also basing the proportional action solely on the process variable. This means that only the integral action responds to changes in the setpoint. The modification to the algorithm does not affect the way the controller responds to process disturbances. Basing proportional action on PV eliminates the instant and possibly very large change in output caused by a sudden change to the setpoint. Depending on the process and tuning this may be beneficial to the response to a setpoint step.

$$\text{MV(t)} = K_p \left(-PV(t) + \frac{1}{T_i} \int_0^t e(\tau)d\tau - T_d \frac{d}{dt}PV(t) \right)$$

Laplace Form of the PID Controller

Sometimes it is useful to write the PID regulator in Laplace transform form:

$$G(s) = K_p + \frac{K_i}{s} + K_d s = \frac{K_d s^2 + K_p s + K_i}{s}$$

Having the PID controller written in Laplace form and having the transfer function of the controlled system makes it easy to determine the closed-loop transfer function of the system.

Series/Interacting Form

Another representation of the PID controller is the series, or *interacting* form,

$$G(s) = K_c \frac{(\tau_i s + 1)}{\tau_i s}(\tau_d s + 1)$$

where the parameters are related to the parameters of the standard form through,

$$K_p = K_c \cdot \alpha, \ T_i = \tau_i \cdot \alpha,$$

and

$$T_d = \frac{\tau_d}{\alpha}$$

with

$$\alpha = 1 + \frac{\tau_d}{\tau_i}.$$

This form essentially consists of a PD and PI controller in series, and it made early (analog) controllers easier to build. When the controllers later became digital, many kept using the interacting form.

Discrete Implementation

The analysis for designing a digital implementation of a PID controller in a microcontroller (MCU) or FPGA device requires the standard form of the PID controller to be *discretized*. Approximations for first-order derivatives are made by backward finite differences. The integral term is discretized, with a sampling time Δt, as follows,

$$\int_0^{t_k} e(\tau)d\tau = \sum_{i=1}^{k} e(t_i)\Delta t$$

The derivative term is approximated as,

$$\frac{de(t_k)}{dt} = \frac{e(t_k) - e(t_{k-1})}{\Delta t}$$

Thus, a *velocity algorithm* for implementation of the discretized PID controller in a MCU is obtained by differentiating $u(t)$, using the numerical definitions of the first and second derivative and solving for $u(t_k)$ and finally obtaining:

$$u(t_k) = u(t_{k-1}) + K_p\left[\left(1 + \frac{\Delta t}{T_i} + \frac{T_d}{\Delta t}\right)e(t_k) + \left(-1 - \frac{2T_d}{\Delta t}\right)e(t_{k-1}) + \frac{T_d}{\Delta t}e(t_{k-2})\right]$$

s.t. $T_i = K_p / K_i, T_d = K_d / K_p$

Pseudocode

Here is a simple software loop that implements a PID algorithm:

```
previous_error = 0

integral = 0

loop:

    error = setpoint - measured_value

    integral = integral + error * dt

    derivative = (error - previous_error) / dt

    output = Kp * error + Ki * integral + Kd * derivative

    previous_error = error

    wait(dt)

    goto loop
```

In this example, two variables that will be maintained within the loop are initialized to zero, then the loop begins. The current *error* is calculated by subtracting the *measured_value* (the process variable, or PV) from the current *setpoint* (SP). Then, *integral* and *derivative* values are calculated, and these and the *error* are combined with three preset gain terms – the proportional gain, the integral gain and the derivative gain – to derive an *output* value.

In the real world, this is D-to-A converted and passed into the process under control as the manipulated variable (MV). The current error is stored elsewhere for re-use in the next differentiation, the program then waits until dt seconds have passed since start, and the loop begins again, reading in new values for the PV and the setpoint and calculating a new value for the error.

Note that for real code, the use of "wait(dt)" might be inappropriate because it doesn't account for time taken by the algorithm itself during the loop, or more importantly, any preemption delaying the algorithm.

Derivative Control

A derivative controller is used in case of "How much time it takes to bring the system to the setpoint"?. This is used wherever transfer lag problems are involved.

It is represented by:

Derivative Controller – proportional to – $\partial \Theta / \partial t$

The derivative controller cannot be used independently. Either it should be used along with proportional controller or with integral controller. The set valve is quickly attained by using the derivative controller.

Derivative action over-correction is good as an experienced operator will give manual over-correction before seeking a final average setting.

Working of Derivative Controller

With an integral controller, we can calculate accumulated error, but with the derivative control, we can calculate the ratio of error change per unit time, hence act as a predictor. Derivative controller action responds to the rate at which the difference between desired value and the measured value is changing that is derivative of the error. Mathematically represented as below,

$$P(t)k = K_D \frac{de_p(t)}{dt}$$

Where,

- KD = Derivative gain

- The derivative controller is also known as Rate action controller or anticipatory controller.

- ep (t)= Desired Value of controlled variable – Measured Value

Applications

The derivative controller is not used alone because it provides no output when the error is constant.

Advantages

- Effect of transportation lag occurred due to the remote allocation of the sensor can be minimized.

- Accumulation of error which will go to affect on actuator saturation can be minimized.

- Peak errors are minimized.

Disadvantages

- Cannot be used when an error is constant. (Derivative of constant value is zero).

- A small change in error will affect largely on controller's output. The high derivative gain will result in heavy overshoots and overall system's stability.

Integral Action

Proportional is just one way to react to an error in the system. The problem with proportional control is that it can't detect trends and adjust to them. This is the job of integral control.

There is another example graph of the error in a system over time on the left of figure. Again, it might be the distance of a robot from an object, or it could be fluid level in a tank, or the temperature in a factory oven. Perhaps the target the robot is following keeps on going away from the robot at a speed that the robot isn't catching up with. Maybe the oven door seal is worn; maybe the fluid draw from the tank is unusually large. Regardless of the cause, since proportional is not designed to react to trends it can't detect and correct the problem. That's where integral control comes into the picture.

Integral measures the area between the error values and the time axis. If the error doesn't return to zero, the area of the error gets larger and larger. The right side of figure shows how the integral output can react to this kind of trend. As the area between the error curve and the time axis increases, the output increases proportional to this area. As a result, the output drives the actuator harder and harder to correct the error.

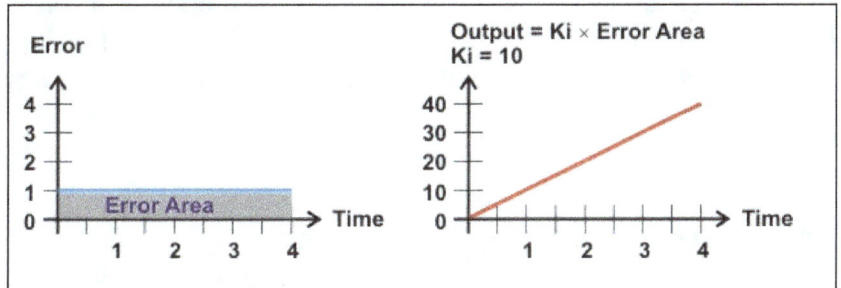

How Integral builds up as error continues to be a problem.

So what happens when the error isn't a straight line, like the curve shown in figure? That's what the calculus operation of integration determines, the area between a curve and an axis. In the case of integral control, as more time passes with an error, the area under the curve grows, and so does the value that the integral calculation will use to drive against the system error. If the error curve drops below the time axis, the buildup of negative area subtracts from the buildup of positive area. When tuned correctly, integral control can help the system hone in on an error of zero.

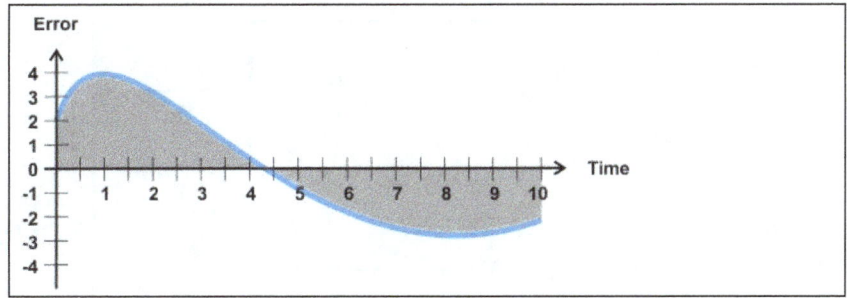

Error Can change with time —so does the area.

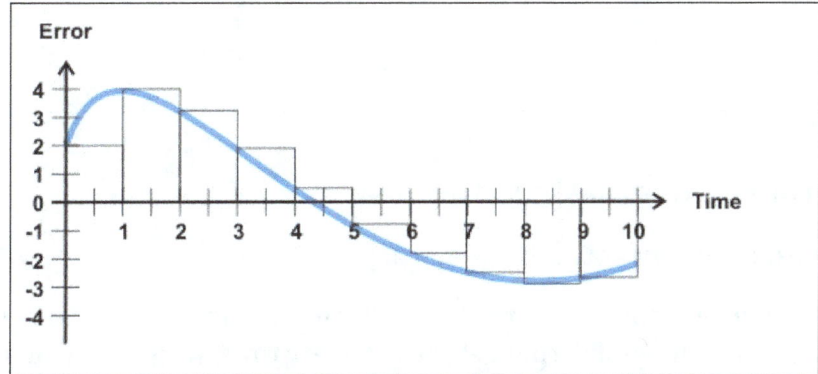

Numerical Integration for Determining Area.

The BASIC Stamp can approximate the error under the curve with numerical integration. Figure shows how you can approximate the error under a curve by adding up the area of a bunch of little rectangles between the error curve and the time axis. The area of each box is the error multiplied by the time between measurements. By adding up all the box areas, you get an approximation of the area under the curve.

So long as your measurements are evenly spaced, you can call the width of each box a value of 1. This makes the math much simpler than trying to account for 20 ms between samples, 5 minutes between samples, or whatever your sampling rate turns out to be. Instead of multiplying error by the time increment between samples and then adding to the next error multiplied by time, you can just multiply each error sample by a time of 1. The result is that you can just keep a running total of error measurements for your integral calculation. Here is an example of how to do this with PBASIC:

'Calculate integral term.

error(Accumulator) = error(Accumulator) + error(Current)

i = Ki * error(Accumulator)

The next example program performs numerical integration on the error signal and adjusts the output accordingly. As with proportional control, there is a constant that scales the integration output to the desired value. For simplicity's sake, we'll use 10 again for Ki. Figure shows a block diagram of the control loop. The term Kp ∫ edt refers to Kp multiplied by the integral of the error over time. In other words, Kp multiplied by the accumulated area between the error curve and the time axis.

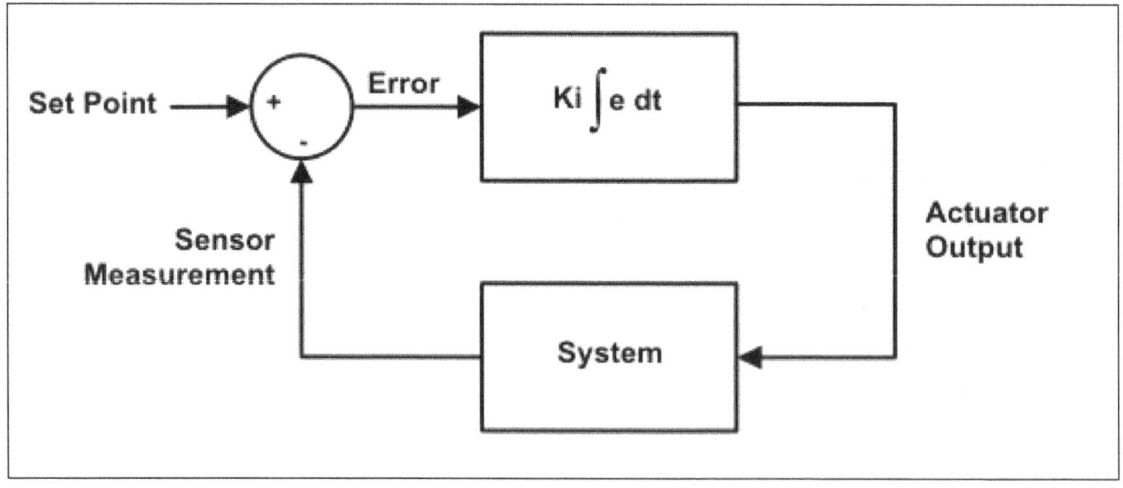

integral control.

Example Program - IntegralAlgorithm.bs2

- Enter, save, and run IntegralAlgorithm.bs2.

- Enter this sequence of values into the Debug Terminal's transmit windowpane: 3 3 3 3 3 3 3 3 3. Notice how the integral output gets larger every time the 3 is repeated. That's integral's job, to detect trends and increase the drive to correct it as needed.

- Now try this sequence: 1 2 3 4 5 4 3 2 1 0 -1 -2 -3 -4 -3 -2 -1 0. The area under this curve is 0 since the negative area is the same as the positive area. That is also what the integral calculation will arrive at when you have finished entering the sequence.

```
' IntegralAlgorithm.bs2
' Demonstrates how integral control influences error correction
' in a feedback loop.


' {$STAMP BS2}
' {$PBASIC 2.5}


SetPoint        CON      0          ' Set point
Ki              CON      10         ' Integral constant


Current         CON      0          ' Array index for current error
Accumulator     CON      1          ' Array index for accumulated error
sensorInput     VAR      Word       ' Input
error           VAR      Word(2)    ' Two element error array
i               VAR      Word       ' Integral term
drive           VAR      Word       ' Output


DO

  DEBUG "Enter sensor input value: "
  DEBUGIN SDEC sensorInput


  ' Calculate error.
  error(Current) = SetPoint - sensorInput


  ' Calculate integral term.
  error(Accumulator) = error(Accumulator) + error(Current)
  i = Ki * error(Accumulator)


  ' Calculate output.
  drive = i
```

```
` Display values.

DEBUG CR, CR, "ERROR", CR,

      SDEC ? SetPoint, SDEC ? sensorInput, SDEC ? error(Current), CR,

      "INTEGRAL", CR,

      SDEC ? Ki, SDEC ? error(accumulator), SDEC ? i, CR,

      "OUTPUT", CR,

      SDEC ? i, SDEC ? drive, CR, CR


    LOOP
```

FUZZY LOGIC

Fuzzy logic is applied with great success in various control application. Almost all the consumer products have fuzzy control. Some of the examples include controlling your room temperature with the help of air-conditioner, anti-braking system used in vehicles, control on traffic lights, washing machines, large economic systems, etc.

The use of Fuzzy Logic in Control Systems

A control system is an arrangement of physical components designed to alter another physical system so that this system exhibits certain desired characteristics. Following are some reasons of using Fuzzy Logic in Control Systems:

- While applying traditional control, one needs to know about the model and the objective function formulated in precise terms. This makes it very difficult to apply in many cases.

- By applying fuzzy logic for control we can utilize the human expertise and experience for designing a controller.

- The fuzzy control rules, basically the IF-THEN rules, can be best utilized in designing a controller.

Assumptions in Fuzzy Logic Control (FLC) Design

While designing fuzzy control system, the following six basic assumptions should be made:

- The plant is observable and controllable: It must be assumed that the input, output as well as state variables are available for observation and controlling purpose.

- Existence of a knowledge body: It must be assumed that there exist a knowledge body having linguistic rules and a set of input-output data set from which rules can be extracted.

- Existence of solution: It must be assumed that there exists a solution.

- 'Good enough' solution is enough: The control engineering must look for 'good enough' solution rather than an optimum one.

- Range of precision: Fuzzy logic controller must be designed within an acceptable range of precision.

- Issues regarding stability and optimality: The issues of stability and optimality must be open in designing Fuzzy logic controller rather than addressed explicitly.

Architecture of Fuzzy Logic Control

The following diagram shows the architecture of Fuzzy Logic Control (FLC).

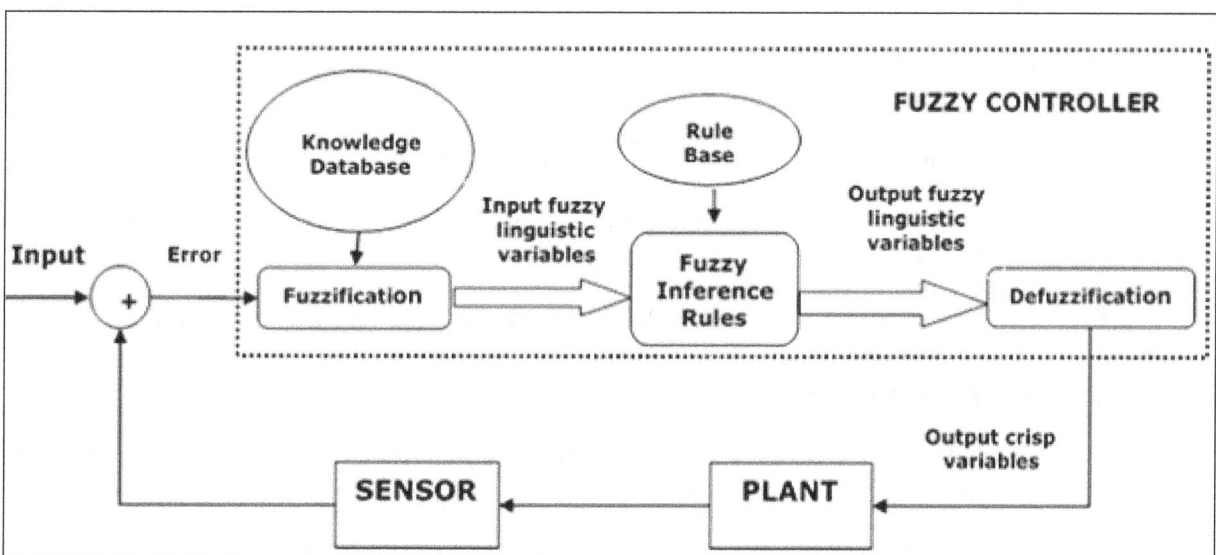

Major Components of FLC

Followings are the major components of the FLC as shown in the figure –

- Fuzzifier: The role of fuzzifier is to convert the crisp input values into fuzzy values.

- Fuzzy Knowledge Base: It stores the knowledge about all the input-output fuzzy relationships. It also has the membership function which defines the input variables to the fuzzy rule base and the output variables to the plant under control.

- Fuzzy Rule Base: It stores the knowledge about the operation of the process of domain.

- Inference Engine: It acts as a kernel of any FLC. Basically it simulates human decisions by performing approximate reasoning.

- Defuzzifier: The role of defuzzifier is to convert the fuzzy values into crisp values getting from fuzzy inference engine.

Steps in Designing FLC

Following are the steps involved in designing FLC –

- Identification of variables: Here, the input, output and state variables must be identified of the plant which is under consideration.

- Fuzzy subset configuration: The universe of information is divided into number of fuzzy subsets and each subset is assigned a linguistic label. Always make sure that these fuzzy subsets include all the elements of universe.

- Obtaining membership function: Now obtain the membership function for each fuzzy subset that we get in the above step.

- Fuzzy rule base configuration: Now formulate the fuzzy rule base by assigning relationship between fuzzy input and output.

- Fuzzification: The fuzzification process is initiated in this step.

- Combining fuzzy outputs: By applying fuzzy approximate reasoning, locate the fuzzy output and merge them.

- Defuzzification: Finally, initiate defuzzification process to form a crisp output.

Advantages of Fuzzy Logic Control

- Cheaper: Developing a FLC is comparatively cheaper than developing model based or other controller in terms of performance.

- Robust: FLCs are more robust than PID controllers because of their capability to cover a huge range of operating conditions.

- Customizable: FLCs are customizable.

- Emulate human deductive thinking: Basically FLC is designed to emulate human deductive thinking, the process people use to infer conclusion from what they know.

- Reliability: FLC is more reliable than conventional control system.

- Efficiency: Fuzzy logic provides more efficiency when applied in control system.

Disadvantages of Fuzzy Logic Control

- Requires lots of data: FLC needs lots of data to be applied.

- Useful in case of moderate historical data: FLC is not useful for programs much smaller or larger than historical data.

- Needs high human expertise: This is one drawback as the accuracy of the system depends on the knowledge and expertise of human beings.

- Needs regular updating of rules: The rules must be updated with time.

References

- Control-systems-introduction, control-systems: tutorialspoint.com, Retrieved 16 May, 2019

- Open-loop-system, systems: electronics-tutorials.ws, Retrieved 04 August, 2019

- Control-system-closed-loop-open-loop-control-system: electrical4u.com, Retrieved 28 June, 2019

- Closed-loop-system, systems: electronics-tutorials.ws, Retrieved 01 May, 2019

- Feedback-control-system-advantages-and-disadvantages, control-systems: electricalacademia.com, Retrieved 19 June, 2019

- Controllers-and-their-characterstics, diy-electronics-devices-54829: brighthubengineering.com, Retrieved 15 August, 2019

- Derivative-controller-control-system: myclassbook.org, Retrieved 03 April, 2019

- Fuzzy-logic-control-system, fuzzy-logic: tutorialspoint.com, Retrieved 24 March, 2019

Mathematical Concepts

Various mathematical concepts are applied within control engineering. Some of these concepts are complex analysis, differential equations and linear hamiltonian control systems. This chapter closely examines these key mathematical concepts of control engineering to provide an extensive understanding of the subject.

COMPLEX ANALYSIS

In the 18th century a far-reaching generalization of analysis was discovered, centred on the so-called imaginary number i = Square root of $\sqrt{-1}$. (In engineering this number is usually denoted by j.) The numbers commonly used in everyday life are known as real numbers, but in one sense this name is misleading. Numbers are abstract concepts, not objects in the physical universe. So mathematicians consider real numbers to be an abstraction on exactly the same logical level as imaginary numbers.

The name *imaginary* arises because squares of real numbers are always positive. In consequence, positive numbers have two distinct square roots—one positive, one negative. Zero has a single square root—namely, zero. And negative numbers have no "real" square roots at all. However, it has proved extremely fruitful and useful to enlarge the number concept to include square roots of negative numbers. The resulting objects are numbers in the sense that arithmetic and algebra can be extended to them in a simple and natural manner; they are imaginary in the sense that their relation to the physical world is less direct than that of the real numbers. Numbers formed by combining real and imaginary components, such as $2 + 3i$, are said to be complex (meaning composed of several parts rather than complicated).

The first indications that complex numbers might prove useful emerged in the 16th century from the solution of certain algebraic equations by the Italian mathematicians Girolamo Cardano and Raphael Bombelli. By the 18th century, after a lengthy and controversial history, they became fully established as sensible mathematical concepts. They remained on the mathematical fringes until it was discovered that analysis, too, can be extended to the complex domain. The result was such a powerful extension of the mathematical tool kit that philosophical questions about the meaning of complex numbers became submerged amid the rush to exploit them. Soon the mathematical community had become so used to complex numbers that it became hard to recall that there had been a philosophical problem at all.

Formal Definition of Complex Numbers

The modern approach is to define a complex number $x + iy$ as a pair of real numbers (x, y) subject to certain algebraic operations. Thus one wishes to add or subtract, $(a, b) \pm (c, d)$, and to multiply,

$(a, b) \times (c, d)$, or divide, $(a, b)/(c, d)$, these quantities. These are inspired by the wish to make $(x, 0)$ behave like the real number x and, crucially, to arrange that $(0, 1)^2 = (-1, 0)$—all the while preserving as many of the rules of algebra as possible. This is a formal way to set up a situation which, in effect, ensures that one may operate with expressions $x + iy$ using all the standard algebraic rules but recalling when necessary that i^2 may be replaced by -1. For example,

$$(1 + 3i)^2 = 1^2 + 2 \cdot 3i + (3i)^2 = 1 + 6i + 9i^2 = 1 + 6i - 9 = -8 + 6i.$$

A geometric interpretation of complex numbers is readily available, inasmuch as a pair (x, y) represents a point in the plane shown in the figure. Whereas real numbers can be described by a single number line, with negative numbers to the left and positive numbers to the right, the complex numbers require a number plane with two axes, real and imaginary.

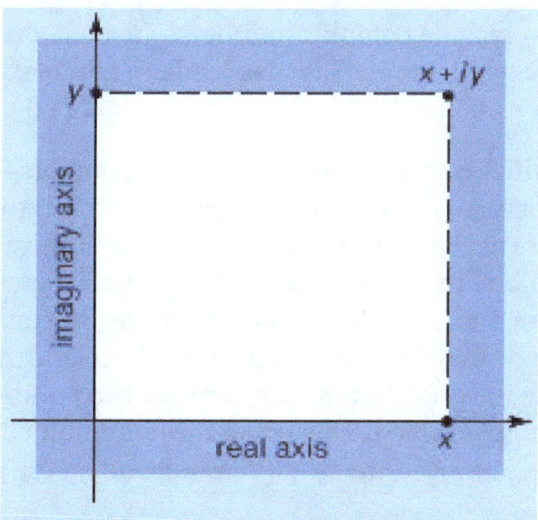

Point in the complex plane.

A point in the complex plane. Unlike real numbers, which can be located by a single signed (positive or negative) number along a number line, complex numbers require a plane with two axes, one axis for the real number component and one axis for the imaginary component. Although the complex plane looks like the ordinary two-dimensional plane, where each point is determined by an ordered pair of real numbers (x, y), the point $x + iy$ is a single number.

Extension of Analytic Concepts to Complex Numbers

Analytic concepts such as limits, derivatives, integrals, and infinite series (all explained in the sections Technical preliminaries and Calculus) are based upon algebraic ideas, together with error estimates that define the limiting process   certain numbers must be arbitrarily well approximated by particular algebraic expressions. In order to represent the concept of an approximation, all that is needed is a well-defined way to measure how "small" a number is. For real numbers this is achieved by using the absolute value $|x|$. Geometrically, it is the distance along the real number line between x and the origin 0. Distances also make sense in the complex plane, and they can be calculated, using Pythagoras's theorem from elementary geometry (the square of the hypotenuse of a right triangle is equal to the sum of the squares of its two sides), by constructing a right triangle such that its hypotenuse spans the distance between two points and its sides are drawn parallel to the coordinate axes. This line of thought leads to the idea that for complex numbers the quantity analogous to $|x|$ is,

$$|x + iy| = \sqrt{x^2 + y^2}.$$

Since all the rules of real algebra extend to complex numbers and the absolute value is defined by an algebraic formula, it follows that analysis also extends to the complex numbers. Formal definitions are taken from the real case, real numbers are replaced by complex numbers, and the real absolute value is replaced by the complex absolute value. Indeed, this is one of the advantages of analytic rigour: without this, it would be far less obvious how to extend such notions as tangent or limit from the real case to the complex.

In a similar vein, the Taylor series for the real exponential and trigonometric functions shows how to extend these definitions to include complex numbers—just use the same series but replace the real variable x by the complex variable z. This idea leads to complex-analytic functions as an extension of real-analytic ones.

Because complex numbers differ in certain ways from real numbers—their structure is simpler in some respects and richer in others—there are differences in detail between real and complex analysis. Complex integration, in particular, has features of complete novelty. A real function must be integrated between limits a and b, and the Riemann integral is defined in terms of a sum involving values spread along the interval from a to b. On the real number line, the only path between two points a and b is the interval whose ends they form. But in the complex plane there are many different paths between two given points. The integral of a function between two points is therefore not defined until a path between the endpoints is specified. This done, the definition of the Riemann integral can be extended to the complex case. However, the result may depend on the path that is chosen.

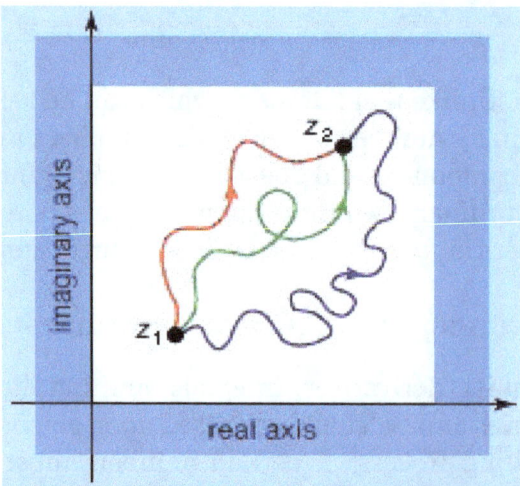

Multiple paths in the complex plane.

Multiple paths in the complex plane. The graph illustrates that two distinct points (z_1 and z_2) in the complex plane may have multiple possible paths between them. Unlike the real number line, where there exists only one path and hence one distance between two points, there are multiple distinct paths between complex numbers. In some cases, this can affect the integral (or length) between two complex points.

Surprisingly, this dependence is very weak. Indeed, sometimes there is no dependence at all. But when there is, the situation becomes extremely interesting. The value of the integral depends only on certain qualitative features of the path—in modern terms, on its topology. (Topology, often characterized as "rubber sheet geometry," studies those properties of a shape that are unchanged if it is continuously deformed by being bent, stretched, and twisted but not torn.) So complex analysis possesses a new ingredient, a kind of flexible geometry, that is totally lacking in real analysis. This gives it a very different flavour.

All this became clear in 1811 when, in a letter to the German astronomer Friedrich Bessel, the German mathematician Carl Friedrich Gauss stated the central theorem of complex analysis:

I affirm now that the integral...has only one value even if taken over different paths, provided [the function]...does not become infinite in the space enclosed by the two paths.

A proof was published by Cauchy in 1825, and this result is now named Cauchy's theorem. Cauchy went on to develop a vast theory of complex analysis and its applications.

Part of the importance of complex analysis is that it is generally better-behaved than real analysis, the many-valued nature of integrals notwithstanding. Problems in the real domain can often be solved by extending them to the complex domain, applying the powerful techniques peculiar to that area, and then restricting the results back to the real domain again. From the mid-19th century onward, the progress of complex analysis was strong and steady. A system of numbers once rejected as impossible and nonsensical led to a powerful and aesthetically satisfying theory with practical applications to aerodynamics, fluid mechanics, electric power generation, and mathematical physics. No area of mathematics has remained untouched by this far-reaching enrichment of the number concept.

Sketched below are some of the key ideas involved in setting up the more elementary parts of complex analysis.

Some Key Ideas of Complex Analysis

A complex number is normally denoted by $z = x + iy$. A complex-valued function f assigns to each z in some region Ω of the complex plane a complex number $w = f(z)$. Usually it is assumed that the region Ω is connected (all in one piece) and open (each point of Ω can be surrounded by a small disk that lies entirely within Ω). Such a function f is differentiable at a point z_0 in Ω if the limit exists as z approaches z_0 of the expression,

$$\frac{f(z) - f(z_0)}{z - z_0}$$

This limit is the derivative $f'(z)$. Unlike real analysis, if a complex function is differentiable in some region, then its derivative is always differentiable in that region, so $f''(z)$ exists. Indeed, derivatives $f^{(n)}(z)$ of all orders $n = 1, 2, 3, \ldots$ exist. Even more strongly, $f(z)$ has a power series expansion $f(z) = c_0 + c_1(z - z_0) + c_2(z - z_0)^2 + \cdots$ with complex coefficients c_j. This series converges for all z lying in some disk with centre z_0. The radius of the largest such disk is called the radius of convergence of the series. Because of this power series representation, a differentiable complex function is said to be analytic.

The elementary functions of real analysis, such as polynomials, trigonometric functions, and exponential functions, can be extended to complex numbers. For example, the exponential of a complex number is defined by,

$$e^z = 1 + z + z^2 / 2! + z^3 / 3! + \cdots$$

where, $n! = n(n-1)\cdots 3 \cdot 2 \cdot 1$. It turns out that the trigonometric functions are related to the exponential by way of Euler's famous formula:

$$e^{i\theta} = \cos(\theta) + i\sin(\theta),$$

which leads to the expressions,

$$\cos(z) = \left(e^{iz} + e^{-iz}\right)/2$$
$$\sin(z) = \left(e^{iz} - e^{-iz}\right)/2i.$$

Every complex number can be written in the form $z = re^{i\theta}$ for real $r \geq 0$ and real θ. Here r is the absolute value (or modulus) of z, and θ is known as its argument. The value of θ is not unique, but the possible values differ only by integer multiples of 2π. In consequence, the complex logarithm is many-valued:

$$\log(z) = \log(re^{i\theta}) = \log|r| + i(\theta + 2n\pi)$$

for any integer n.

The integral,

$$\int_C f(z)dz$$

of an analytic function f along a curve (or contour) C in the complex plane is defined in a similar manner to the real Riemann integral. Cauchy's theorem, mentioned above, states that the value of such an integral is the same for two contours C_1 and C_2, provided both curves lie inside a simply connected region Ω—a region with no "holes." When Ω has holes, the value of the integral depends on the topology of the curve C but not its precise form. The essential feature is how many times C winds around a given hole—a number that is related to the many-valued nature of the complex logarithm.

DIFFERENTIAL EQUATIONS

Differential equation model is a time domain mathematical model of control systems. Follow these steps for differential equation model.

- Apply basic laws to the given control system.

- Get the differential equation in terms of input and output by eliminating the intermediate variables.

Example:

Consider the following electrical system as shown in the following figure. This circuit consists of resistor, inductor and capacitor. All these electrical elements are connected in series. The input voltage applied to this circuit is v_i and the voltage across the capacitor is the output voltage v_o.

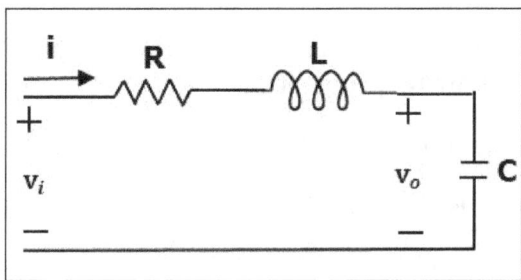

Mesh equation for this circuit is

$$v_i = Ri + L\frac{di}{dt} + v_o$$

Substitute, the current passing through capacitor $i = c\frac{dv_o}{dt}$ in the above equation.

$$\Rightarrow v_i = RC\frac{dv_o}{dt} + LC\frac{d^2v_o}{dt^2} + v_o$$

$$\Rightarrow \frac{d^2v_o}{dt^2} + \left(\frac{R}{L}\right)\frac{dv_o}{dt} + \left(\frac{1}{LC}\right)v_o = \left(\frac{1}{LC}\right)v_i$$

The above equation is a second order differential equation.

LINEAR HAMILTONIAN CONTROL SYSTEMS

Linear Hamiltonian System are the systems that are that are linear differential equations. For this kind of systems the linear algebra is the cornerstone of many of the results that can be obtained.

Controlling Hamiltonian systems has recently attracted the attention of researchers due to their applications in areas such as quantum control satellite control, mixing control or power system control.

Preliminaries

Hamiltonian Matrices

Let J be a real skew-symmetric matrix $2n \times 2n$ defined on the form

$$J = \begin{pmatrix} 0_n & I_n \\ -I_n & 0_n \end{pmatrix}$$

where O_n and I_n are zero and identity matrices

Notice that $J^t = J^{-1} = -J$

A real $2n \times 2n$ matrix A, is said a Hamiltonian matrix if the matrix JA is symmetric

Attending at definition and J matrix properties, an well known equivalent definition of Hamiltonian character is given in the following proposition

Proposition: The matrix A is Hamiltonian if and only if, it verifies the equation $A^t J + JA = 0$

Proof: JA is symmetric if and only if $JA = (JA)^t$

$$JA = (JA)^t \Leftrightarrow JA - (JA)^t = 0$$

$$JA - A^t J^t = 0 \Leftrightarrow A^t J + JA = 0$$

That is to say, A is a Hamiltonian matrix $\Leftrightarrow A^t J + JA = 0$

The set of $2n \times 2n$ Hamiltonian matrices is expressed by

$$\mathcal{H}^n = \{A \in \mathcal{M}(\mathbb{R})_{2n \times 2n} \mid A^t J + JA = 0\}$$

Properties: Let A be a Hamiltonian matrix.

i) Suppose A written as a block matrix in the form

$$A = \begin{pmatrix} A_{11} & A_{12} \\ A_{21} & A_{22} \end{pmatrix}$$

where, A_{11}, A_{12}, A_{21}, and A_{22} are n-order square matrices. Then, the matrices A_{12} and A_{21} are symmetric, and $A_{11} + A_{22}^t = 0$.

ii) $A = JS$ where S is a symmetric matrix. A^t is Hamiltonian.

iii) The sum (and any linear combination) of two Hamiltonian matrices is also Hamiltonian, as well as is their commutator.

iv) The space of all Hamiltonian matrices is a Lie algebra, denoted $sp(2n)$. The dimension of $sp(2n)$ is $2n^2 + n$. The corresponding Lie group is the symplectic group $sp(2n)$. This group consists of the symplectic matrices, that is to say the set of matrices A which satisfy $A^t JA = J$.

Dynamical systems

Given a dynamical system with multiples inputs $u_1(t) \dots u_m(t)$, multiples outputs $y_1(t) \dots y_p(t)$ and $x_1(t) \dots x_n(t)$ state variables, it is possible model its behaviour with n first order differential equations.

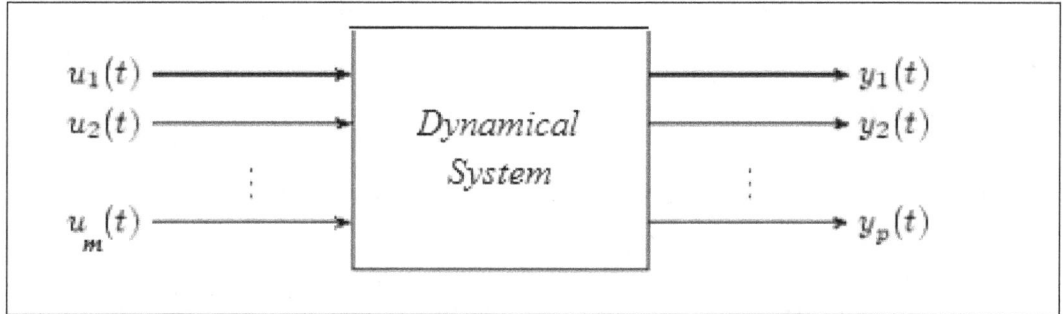

Dynamical system with multiples inputs and outputs.

$$\begin{cases} \dot{x}_1 = f_1\big(x_1(t)\ldots x_n(t), u_1(t)\ldots u_m(t), t\big) \\ \qquad\qquad \vdots \\ \dot{x}_n = f_n\big(x_1(t)\ldots x_n(t), u_1(t)\ldots u_m(t), t\big) \end{cases}$$

and with the p output equations,

$$\begin{cases} y_1 = g_1\big(x_1(t)\ldots x_n(t), u_1(t)\ldots u_m(t), t\big) \\ \qquad\qquad \vdots \\ y_p = g_p\big(x_1(t)\ldots x_n(t), u_1(t)\ldots u_m(t), t\big) \end{cases}$$

Taking,

$$u(t) = \begin{pmatrix} u_1(t) \\ u_2(t) \\ \vdots \\ u_m(t) \end{pmatrix};\ x(t) = \begin{pmatrix} x_1(t) \\ x_2(t) \\ \vdots \\ x_n(t) \end{pmatrix};\ y(t) = \begin{pmatrix} y_1(t) \\ y_2(t) \\ \vdots \\ y_p(t) \end{pmatrix}$$

and rewriting in vectorial form,

$$\dot{x}(t) = f\big(x(t), u(t), t\big)$$
$$y(t) = g\big(x(t), u(t), t\big)$$

where,

$$\dot{x}(t) = \frac{d}{dt}x(t)$$

it is obtained a general form to model dynamical system.

Characterization

At practice, the most of dynamical systems works like linear dynamical systems. And if not, in any case, it becomes necessary linearize them by zones to study. This is the reason for the interest in using a linear algebra techniques to study them.

So, the differential state equations most used to describe the behaviour of a system are,

$$\dot{x}(t) = Ax(t) + Bu(t)$$
$$y(t) = Cx(t) + Du(t)$$

where $x(t) \in \mathbb{R}^n$ is called the state vector, $u(t) \in \mathbb{R}^m$ the input vector, $t \geq 0$, A, B, C and D are real matrices of appropriate sizes.

The most typical representation of a system modelled by $\begin{matrix} \dot{x}(t) = Ax(t) + Bu(t) \\ y(t) = Cx(t) + Du(t) \end{matrix}$, valid for all linear system, is using a blocks diagram as shown in the figure.

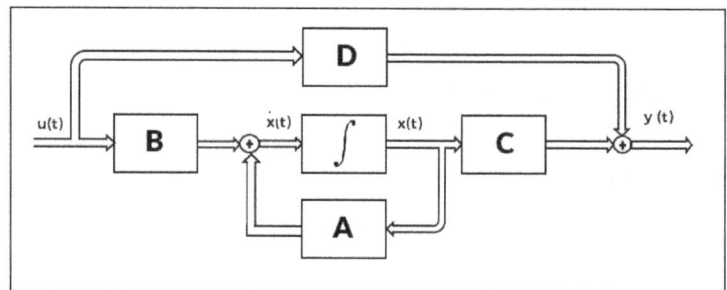

Blocks diagram of an open loop dynamical system.

It is important to note that $\begin{matrix} \dot{x}(t) = Ax(t) + Bu(t) \\ y(t) = Cx(t) + Du(t) \end{matrix}$ refers to linear timeinvariant systems. By abuse of language, they only will be referred as linear systems while they make no mistake about it.

Linear Hamiltonian Systems

The system $\begin{matrix} \dot{x}(t) = Ax(t) + Bu(t) \\ y(t) = Cx(t) + Du(t) \end{matrix}$ is called linear Hamiltonian if and only if is A is a Hamiltonian matrix.

In case of Linear Hamiltonian systems, the own Hamiltonian determines concretes sizes for matrices and vectors involved in the model. So,

Corollary: Let us consider a linear Hamiltonian system given by $\begin{matrix} \dot{x}(t) = Ax(t) + Bu(t) \\ y(t) = Cx(t) + Du(t) \end{matrix}$ then,

$$A \in \mathcal{H}^n$$

$$B \in \mathcal{M}(\mathbb{R})_{2n \times m}$$

$$C \in \mathcal{M}(\mathbb{R})_{p \times 2n}$$

$$D \in \mathcal{M}(\mathbb{R})_{p \times m}$$

It is important to note and remember that m were the numbers of inputs, p the numbers of outputs and n the minimum number of states to describe the system.

Control of Hamiltonian Linear systems

Roughly speaking, controllability denotes the ability to move a system around in its entire configuration space using only certain admissible manipulations. More specifically, the system

$$\dot{x}(t) = Ax(t) + Bu(t)$$
$$y(t) = Cx(t) + Du(t)$$

is controllable if and only if the controllability matrix has full row rank:

$$\text{rank} \left(B \ AB \ldots \ ^{n-1}B \right) = n$$

When the system is controllable, there exists a feedback K in such a way that the resultant closed loop system,

$$\dot{x}(t) = (A - BK)x(t)$$
$$y(t) = (C - DK)x(t)$$

have a desired stable solution.

Given an open loop linear Hamiltonian, the interest of the study concerns in obtain the feedback making the closes loop system stable reaching the intended states and preserving the Hamiltonian structure of the system.

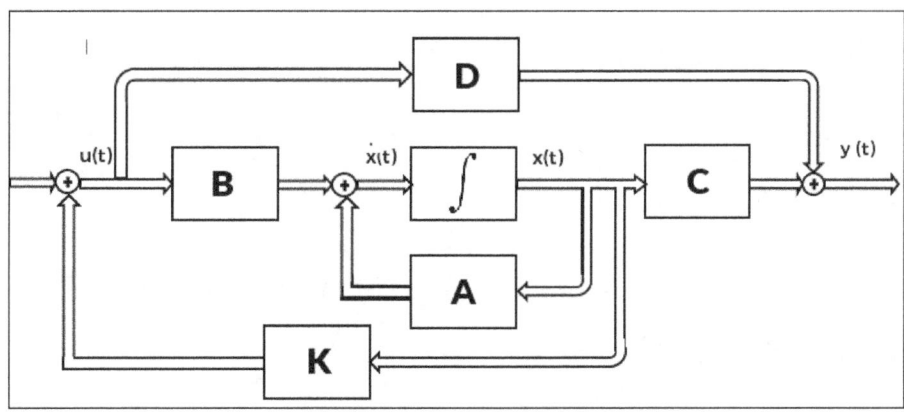

Closed loop Linear system.

So, the system in closed loop $\dfrac{\dot{x}(t) = (A - BK)x(t)}{y(t) = (C - DK)x(t)}$ will be Linear Hamiltonian if and only if the

system matrix is Hamiltonian, i.e. $A - BK \in \mathcal{H}^n$

Proposition: Given a Linear Hamiltonian system in open loop, the closed loop system will continue being Hamiltonian if and only if BK is a Hamiltonian matrix.

Theorem: Let consider a Linear Hamiltonian System. The matrix $K = \left(K_1 \ K_2 \right)$ is a feedback matrix preserving the hamiltonian character if and only if K is a solution of the Sylvester generalized equation,

$$K_1^t B_1^t + B_2 K_2 = 0$$

in such away tat $B_1 K_2$ and $B_2 K_1$ are symmetrical matrices,

Proof: It suffices to compute,

$$M = \begin{pmatrix} 0_n & I_n \\ -I_n & 0_n \end{pmatrix} \begin{pmatrix} B_1 \\ B_2 \end{pmatrix} (K_1 \ K_2)$$

and force it to be symmetrical: $M = M^t$.

Using Kronecker product ($\otimes$) and vectorializing operator (vec) the following corollary is obtained.

Corollary: Let consider a linear Hamiltonian system. The matrix $K = (K_1 \ K_2)$ is a feedback matrix preserving the hamiltonian character if and only if K is a solution of the following linear system

$$\left(B_1^t \otimes I_n \ I_n \otimes B_2 \right) \begin{pmatrix} \text{vec } K_1^t \\ \text{vec } K_2 \end{pmatrix}$$

in such away tat $B_1 K_2$ and $B_2 K_1$ are symmetrical matrices

Proof: It suffices to remember, for more information), that If $A = (a_j^i) \in M_{n \times m}(C)$ and $A = (a_j^i) \in M_{n \times m}(C) \ B \in M_{p \times q}(C)$ then,

$$A \otimes B \begin{pmatrix} a_1^1 B & a_2^1 B & \dots & a_m^1 B \\ a_1^2 B & a_2^2 B & \dots & a_m^2 B \\ \vdots & \vdots & & \vdots \\ a_1^n B & a_2^n B & \dots & a_m^n B \end{pmatrix} \in M_{np \times mq}(C).$$

If $X = (x_j^i) \in M_{n \times m}(C)$, denoting $x_i = (x_1^i, \dots, x_m^i)$ the i-th row of the matrix X.

$$vec : M_{n \times m}(C) \to M_{nm \times 1}(C)$$

$$X \to (x_1 \ x_2 \ \dots \ x_n)^t.$$

If $A \in M_{n \times m}(C)$, $X \in M_{m \times p}(C)$ and $B \in M_{p \times q}(C)$ then, vec $(AX \ B) = (B^t \otimes A)$ vec (X).

Particular Cases

Suppose now that $B_1 = 0$ then, $B_2 K_2 = 0$ and $B_2 K_1$ being a symmetric matrix.

Analogously, if $B_2 = 0$ then, $B_1 K_1 = 0$ and $B_1 K_2$ being a symmetric matrix.

If ($B_1 = B_2$ then, $B_1^t \otimes I_n \ I_n \otimes B_1$)

$$\begin{pmatrix} \text{vec } K_1^t \\ K_2^t \end{pmatrix} = 0 \text{ with } B_1 K_1 \text{ and } B_1 K_1 \text{ symmetric matrices.}$$

Example: Let B be a matrix with $B_1 = 0$ and $B_2 = \begin{pmatrix} 1 & 2 \\ 3 & 4 \end{pmatrix}$.

Then, $K_2 = 0$ verifies $B_2 K_2 = 0$) and

taking $K_1 = \begin{pmatrix} x & y \\ z & t \end{pmatrix}$ with $y + 2t = 3x + 4z$ $B_2 K_1$ is a symmetric matrix, and BK is a Hamiltonian matrix:

$$\begin{pmatrix} 0 & 0 \\ 0 & 0 \\ 1 & 2 \\ 3 & 4 \end{pmatrix} \begin{pmatrix} a & 3a+4c-2d & 0 & 0 \\ c & d & 0 & 0 \end{pmatrix} =$$

$$\begin{pmatrix} 0 & 0 & 0 & 0 \\ 0 & 0 & 0 & 0 \\ a+2c & 3a+4c & 0 & 0 \\ 3a+4c & 9a+12c-2d & 0 & 0 \end{pmatrix}$$

and

$$\begin{pmatrix} 0 & 0 & 1 & 0 \\ 0 & 0 & 0 & 1 \\ -1 & 0 & 0 & 0 \\ 0 & -1 & 0 & 0 \end{pmatrix} \begin{pmatrix} 0 & 0 & 0 & 0 \\ 0 & 0 & 0 & 0 \\ a+2c & 3a+4c & 0 & 0 \\ 3a+4c & 9a+12c-2d & 0 & 0 \end{pmatrix} =$$

$$\begin{pmatrix} a+2c & 3a+4c & 0 & 0 \\ 3a+4c & 9a+12c-2d & 0 & 0 \\ 0 & 0 & 0 & 0 \\ 0 & 0 & 0 & 0 \end{pmatrix}$$

Applications

Control engineering has a wide range of applications in many modern automobiles. A few of these applications include aircraft flight control system, control loading system, electronic flight instrument system, fire-control system, intelligent flight control system, guidance system, etc. The topics elaborated in this chapter will help in gaining a better perspective about these applications of control engineering.

AIRCRAFT FLIGHT CONTROL SYSTEM

As aviation matured and aircraft designers learned more about aerodynamics, the industry produced larger and faster aircraft. Therefore, the aerodynamic forces acting upon the control surfaces increased exponentially. To make the control force required by pilots manageable, aircraft engineers designed more complex systems. At first, hydromechanical designs, consisting of a mechanical circuit and a hydraulic circuit, were used to reduce the complexity, weight, and limitations of mechanical flight controls systems.

As aircraft became more sophisticated, the control surfaces were actuated by electric motors, digital computers, or fiber optic cables. Called "fly-by-wire," this flight control system replaces the physical connection between pilot controls and the flight control surfaces with an electrical interface. In addition, in some large and fast aircraft, controls are boosted by hydraulically or electrically actuated systems. In both the fly-by-wire and boosted controls, the feel of the control reaction is fed back to the pilot by simulated means.

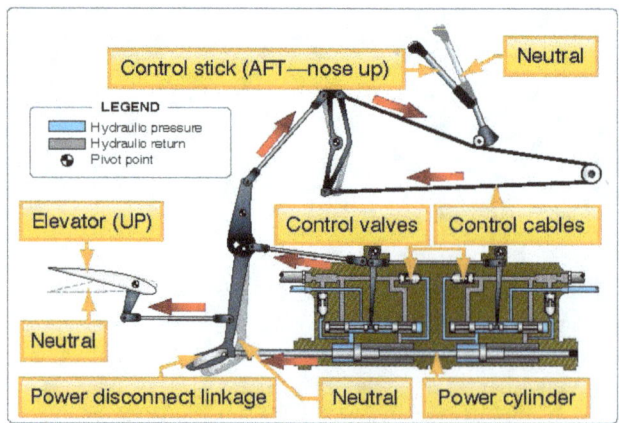

Hydromechanical flight control system.

Current research at the National Aeronautics and Space Administration (NASA) Dryden Flight

Research Center involves Intelligent Flight Control Systems (IFCS). The goal of this project is to develop an adaptive neural network-based flight control system. Applied directly to flight control system feedback errors, IFCS provides adjustments to improve aircraft performance in normal flight, as well as with system failures. With IFCS, a pilot is able to maintain control and safely land an aircraft that has suffered a failure to a control surface or damage to the airframe. It also improves mission capability, increases the reliability and safety of flight, and eases the pilot workload.

Today's aircraft employ a variety of flight control systems. For example, some aircraft in the sport pilot category rely on weight-shift control to fly while balloons use a standard burn technique. Helicopters utilize a cyclic to tilt the rotor in the desired direction along with a collective to manipulate rotor pitch and anti-torque pedals to control yaw.

For additional information on flight control systems, refer to the appropriate handbook for information related to the flight control systems and characteristics of specific types of aircraft.

Flight Control Systems

Flight Controls

Aircraft flight control systems consist of primary and secondary systems. The ailerons, elevator (or stabilator), and rudder constitute the primary control system and are required to control an aircraft safely during flight. Wing flaps, leading edge devices, spoilers, and trim systems constitute the secondary control system and improve the performance characteristics of the airplane or relieve the pilot of excessive control forces.

Primary Flight Controls

Aircraft control systems are carefully designed to provide adequate responsiveness to control inputs while allowing a natural feel. At low airspeeds, the controls usually feel soft and sluggish, and the aircraft responds slowly to control applications. At higher airspeeds, the controls become increasingly firm and aircraft response is more rapid.

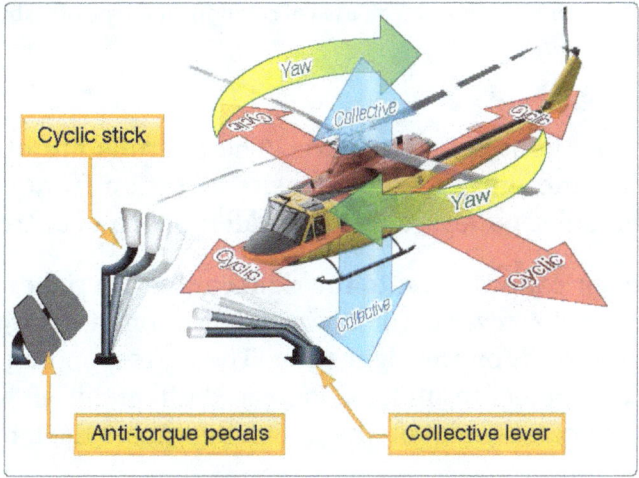

Helicopter flight control system.

Movement of any of the three primary flight control surfaces (ailerons, elevator or stabilator, or rudder), changes the airflow and pressure distribution over and around the airfoil. These changes affect the lift and drag produced by the airfoil/ control surface combination, and allow a pilot to control the aircraft about its three axes of rotation.

Design features limit the amount of deflection of flight control surfaces. For example, control-stop mechanisms may be incorporated into the flight control linkages, or movement of the control column and rudder pedals may be limited. The purpose of these design limits is to prevent the pilot from inadvertently overcontrolling and overstressing the aircraft during normal maneuvers.

A properly designed aircraft is stable and easily controlled during normal maneuvering. Control surface inputs cause movement about the three axes of rotation. The types of stability an aircraft exhibits also relate to the three axes of rotation.

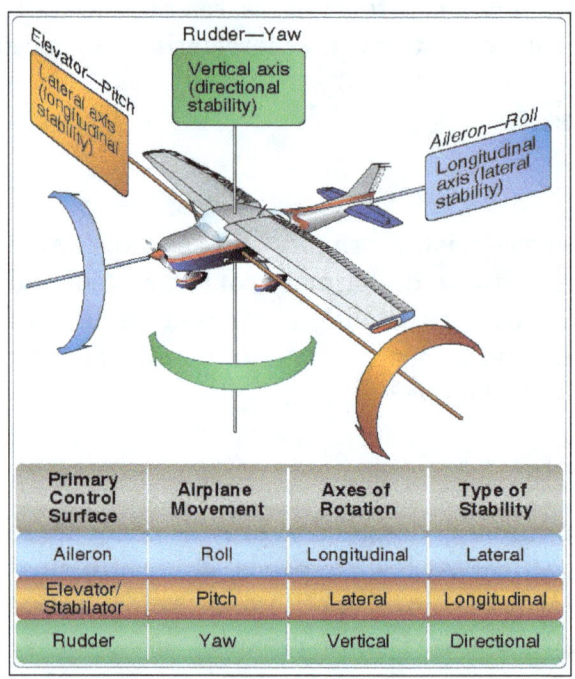

Primary Control Surface	Airplane Movement	Axes of Rotation	Type of Stability
Aileron	Roll	Longitudinal	Lateral
Elevator/ Stabilator	Pitch	Lateral	Longitudinal
Rudder	Yaw	Vertical	Directional

Airplane controls, movement, axes of rotation, and type of stability.

Ailerons

Ailerons control roll about the longitudinal axis. The ailerons are attached to the outboard trailing edge of each wing and move in the opposite direction from each other. Ailerons are connected by cables, bellcranks, pulleys, and push-pull tubes to a control wheel or control stick.

Moving the control wheel, or control stick, to the right causes the right aileron to deflect upward and the left aileron to deflect downward. The upward deflection of the right aileron decreases the camber resulting in decreased lift on the right wing. The corresponding downward deflection of the left aileron increases the camber resulting in increased lift on the left wing. Thus, the increased lift on the left wing and the decreased lift on the right wing causes the aircraft to roll to the right.

Adverse Yaw

Since the downward deflected aileron produces more lift as evidenced by the wing raising, it also produces more drag. This added drag causes the wing to slow down slightly. This results in the aircraft yawing toward the wing which had experienced an increase in lift (and drag). From the pilot's perspective, the yaw is opposite the direction of the bank. The adverse yaw is a result of differential drag and the slight difference in the velocity of the left and right wings.

Adverse yaw becomes more pronounced at low airspeeds. At these slower airspeeds, aerodynamic pressure on control surfaces are low, and larger control inputs are required to effectively maneuver the aircraft. As a result, the increase in aileron deflection causes an increase in adverse yaw. The yaw is especially evident in aircraft with long wing spans.

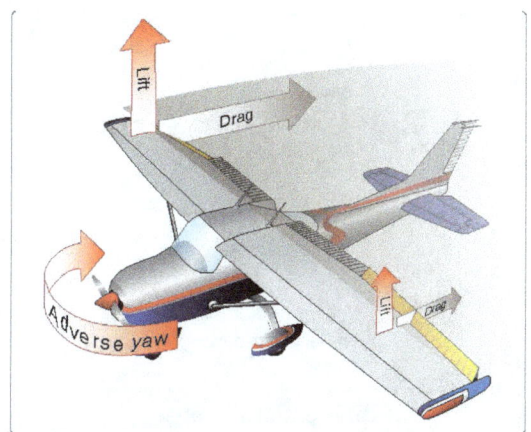

Adverse yaw is caused by higher drag on the outside wing that is producing more lift.

Application of the rudder is used to counteract adverse yaw. The amount of rudder control required is greatest at low airspeeds, high angles of attack, and with large aileron deflections. Like all control surfaces at lower airspeeds, the vertical stabilizer/rudder becomes less effective and magnifies the control problems associated with adverse yaw.

All turns are coordinated by use of ailerons, rudder, and elevator. Applying aileron pressure is necessary to place the aircraft in the desired angle of bank, while simultaneous application of rudder pressure is necessary to counteract the resultant adverse yaw. Additionally, because more lift is required during a turn than during straight-and-level flight, the angle of attack (AOA) must be increased by applying elevator back pressure. The steeper the turn, the more elevator back pressure that is needed.

As the desired angle of bank is established, aileron and rudder pressures should be relaxed. This stops the angle of bank from increasing, because the aileron and rudder control surfaces are in a neutral and streamlined position. Elevator back pressure should be held constant to maintain altitude. The roll-out from a turn is similar to the roll-in, except the flight controls are applied in the opposite direction. The aileron and rudder are applied in the direction of the roll-out or toward the high wing. As the angle of bank decreases, the elevator back pressure should be relaxed as necessary to maintain altitude.

In an attempt to reduce the effects of adverse yaw, manufacturers have engineered four systems: differential ailerons, frise-type ailerons, coupled ailerons and rudder, and flaperons.

Differential Ailerons

With differential ailerons, one aileron is raised a greater distance than the other aileron and is lowered for a given movement of the control wheel or control stick. This produces an increase in drag on the descending wing. The greater drag results from deflecting the up aileron on the descending wing to a greater angle than the down aileron on the rising wing. While adverse yaw is reduced, it is not eliminated completely.

Frise-Type Ailerons

With a frise-type aileron, when pressure is applied to the control wheel, or control stick, the aileron that is being raised pivots on an offset hinge. This projects the leading edge of the aileron into the airflow and creates drag. It helps equalize the drag created by the lowered aileron on the opposite wing and reduces adverse yaw.

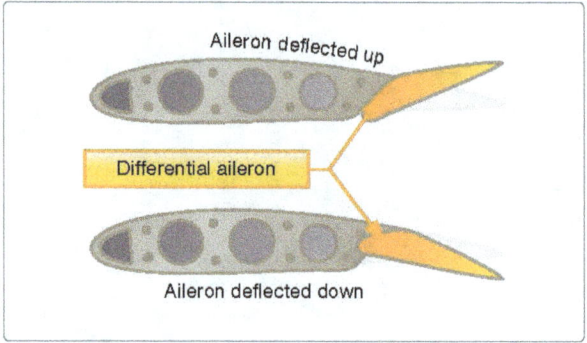

Differential ailerons.

The frise-type aileron also forms a slot so air flows smoothly over the lowered aileron, making it more effective at high angles of attack. Frise-type ailerons may also be designed to function differentially. Like the differential aileron, the frise-type aileron does not eliminate adverse yaw entirely. Coordinated rudder application is still needed when ailerons are applied.

Coupled Ailerons and Rudder

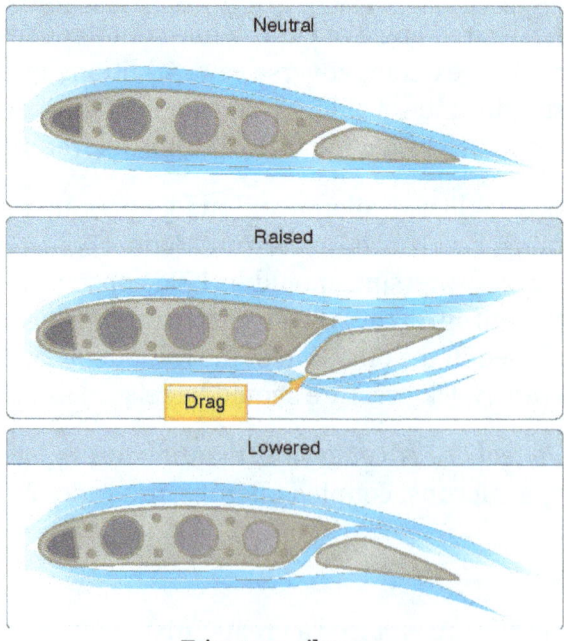

Frise-type ailerons.

Coupled ailerons and rudder are linked controls. This is accomplished with rudder-aileron interconnect springs, which help correct for aileron drag by automatically deflecting the rudder at the same time the ailerons are deflected. For example, when the control wheel, or control stick, is moved to produce a left roll, the interconnect cable and spring pulls forward on the left rudder pedal just enough to prevent the nose of the aircraft from yawing to the right. The force applied to the rudder by the springs can be overridden if it becomes necessary to slip the aircraft.

Flaperons

Flaperons combine both aspects of flaps and ailerons. In addition to controlling the bank angle of an aircraft like conventional ailerons, flaperons can be lowered together to function much the same as a dedicated set of flaps. The pilot retains separate controls for ailerons and flaps. A mixer is used to combine the separate pilot inputs into this single set of control surfaces called flaperons. Many designs that incorporate flaperons mount the control surfaces away from the wing to provide undisturbed airflow at high angles of attack and low airspeeds.

Elevator

The elevator controls pitch about the lateral axis. Like the ailerons on small aircraft, the elevator is connected to the control column in the flight deck by a series of mechanical linkages. Aft movement of the control column deflects the trailing edge of the elevator surface up. This is usually referred to as the up-elevator position.

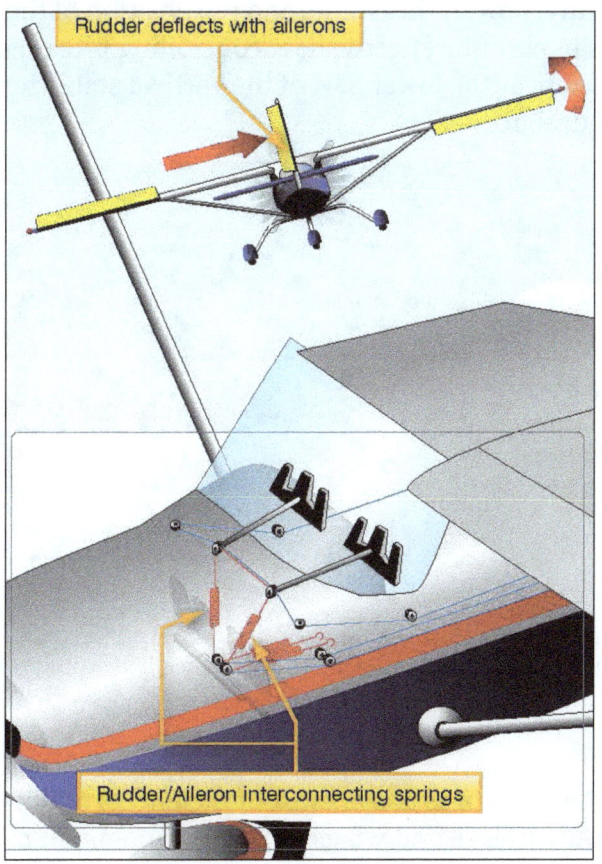

Coupled ailerons and rudder.

Flaperons on a Skystar Kitfox MK 7.

The up-elevator position decreases the camber of the elevator and creates a downward aerodynamic force, which is greater than the normal tail-down force that exists in straight-andlevel flight. The overall effect causes the tail of the aircraft to move down and the nose to pitch up. The pitching moment occurs about the center of gravity (CG). The strength of the pitching moment is determined by the distance between the CG and the horizontal tail surface, as well as by the aerodynamic effectiveness of the horizontal tail surface. Moving the control column forward has the opposite effect. In this case, elevator camber increases, creating more lift (less tail-down force) on the horizontal stabilizer/elevator. This moves the tail upward and pitches the nose down. Again, the pitching moment occurs about the CG.

As mentioned earlier, stability, power, thrustline, and the position of the horizontal tail surfaces on the empennage are factors in elevator effectiveness controlling pitch. For example, the horizontal tail surfaces may be attached near the lower part of the vertical stabilizer, at the midpoint, or at the high point, as in the T-tail design.

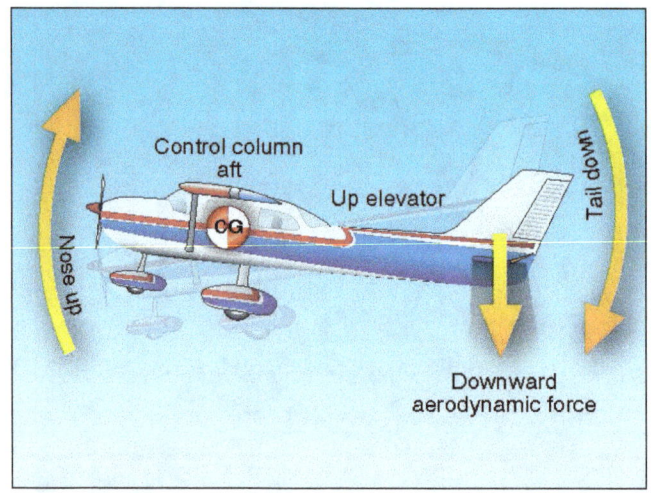

The elevator is the primary control for changing the pitch attitude of an aircraft.

T-Tail

In a T-tail configuration, the elevator is above most of the effects of downwash from the propeller, as well as airflow around the fuselage and wings during normal flight conditions. Operation of the elevators in this undisturbed air allows control movements that are consistent throughout most flight regimes. T-tail designs have become popular on many light and large aircraft, especially those with aft fuselagemounted engines because the T-tail configuration removes the tail from the exhaust blast of the engines. Seaplanes and amphibians often have T-tails in order to keep the horizontal surfaces as far from the water as possible. An additional benefit is reduced noise and vibration inside the aircraft.

In comparison with conventional-tail aircraft, the elevator on a T-tail aircraft must be moved a greater distance to raise the nose a given amount when traveling at slow speeds. This is because the conventional-tail aircraft has the downwash from the propeller pushing down on the tail to assist in raising the nose.

Aircraft controls are rigged so that an increase in control force is required to increase control travel. The forces required to raise the nose of a T-tail aircraft are greater than the forces required to raise the nose of a conventional-tail aircraft. Longitudinal stability of a trimmed aircraft is the same for both types of configuration, but the pilot must be aware that the required control forces are greater at slow speeds during takeoffs, landings, or stalls than for similar size aircraft equipped with conventional tails.

T-tail aircraft also require additional design considerations to counter the problem of flutter. Since the weight of the horizontal surfaces is at the top of the vertical stabilizer, the moment arm created causes high loads on the vertical stabilizer that can result in flutter. Engineers must compensate for this by increasing the design stiffness of the vertical stabilizer, usually resulting in a weight penalty over conventional tail designs.

When flying at a very high AOA with a low airspeed and an aft CG, the T-tail aircraft may be more susceptible to a deep stall. In this condition, the wake of the wing impinges on the tail surface and renders it almost ineffective. The wing, if fully stalled, allows its airflow to separate right after the leading edge. The wide wake of decelerated, turbulent air blankets the horizontal tail and hence its effectiveness diminished significantly. In these circumstances, elevator or stabilator control is reduced (or perhaps eliminated) making it difficult to recover from the stall. It should be noted that an aft CG is often a contributing factor in these incidents, since similar recovery problems are also found with conventional tail aircraft with an aft CG. Deep stalls can occur on any aircraft but are more likely to occur on aircraft with "T" tails as a high AOA may be more likely to place the wings separated airflow into the path of the horizontal surface of the tail. Additionally, the distance between the wings and the tail, the position of the engines (such as being mounted on the tail) may increase the susceptibility of deep stall events. Therefore a deep stall may be more prevalent on transport versus general aviation aircraft.

Since flight at a high AOA with a low airspeed and an aft CG position can be dangerous, many aircraft have systems to compensate for this situation. The systems range from control stops to elevator down springs. On transport category jets, stick pushers are commonly used. An elevator down spring assists in lowering the nose of the aircraft to prevent a stall caused by the aft CG position. The stall occurs because the properly trimmed airplane is flying with the elevator in a trailing edge down position, forcing the tail up and the nose down. In this unstable condition, if the aircraft encounters turbulence and slows down further, the trim tab no longer positions the elevator in the nosedown position. The elevator then streamlines, and the nose of the aircraft pitches upward, possibly resulting in a stall.

The elevator down spring produces a mechanical load on the elevator, causing it to move toward the nose-down position if not otherwise balanced. The elevator trim tab balances the elevator down spring to position the elevator in a trimmed position. When the trim tab becomes ineffective, the down spring drives the elevator to a nose-down position. The nose of the aircraft lowers, speed builds up, and a stall is prevented.

The elevator must also have sufficient authority to hold the nose of the aircraft up during the roundout for a landing. In this case, a forward CG may cause a problem. During the landing flare, power is usually reduced, which decreases the airflow over the empennage. This, coupled with the reduced landing speed, makes the elevator less effective.

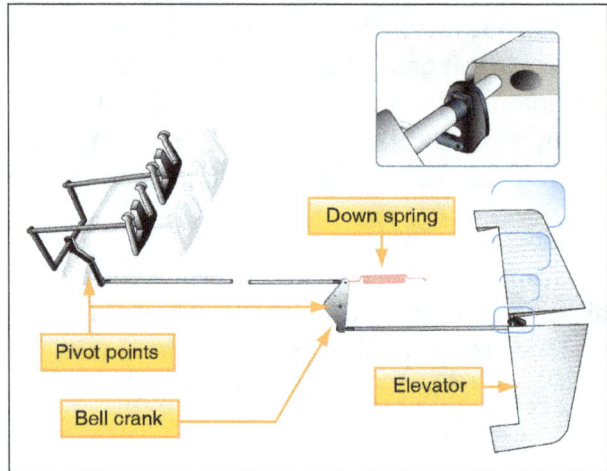

When the aerodynamic efficiency of the horizontal tail surface is inadequate due to an aft CG condition, an elevator down spring may be used to supply a mechanical load to lower the nose.

Stabilator

As mentioned in Chapter 3, Aircraft Structure, a stabilator is essentially a one-piece horizontal stabilizer that pivots from a central hinge point. When the control column is pulled back, it raises the stabilator's trailing edge, pulling the nose of the aircraft. Pushing the control column forward lowers the trailing edge of the stabilator and pitches the nose of the aircraft down.

Because stabilators pivot around a central hinge point, they are extremely sensitive to control inputs and aerodynamic loads. Antiservo tabs are incorporated on the trailing edge to decrease sensitivity. They deflect in the same direction as the stabilator. This results in an increase in the force required to move the stabilator, thus making it less prone to pilot-induced overcontrolling. In addition, a balance weight is usually incorporated in front of the main spar. The balance weight may project into the empennage or may be incorporated on the forward portion of the stabilator tips.

Canard

The canard design utilizes the concept of two lifting surfaces. The canard functions as a horizontal stabilizer located in front of the main wings. In effect, the canard is an airfoil similar to the horizontal surface on a conventional aft-tail design. The difference is that the canard actually creates lift and holds the nose up, as opposed to the aft-tail design which exerts downward force on the tail to prevent the nose from rotating downward.

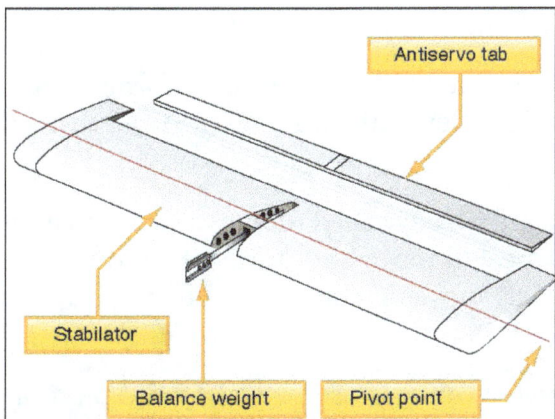

The stabilator is a one-piece horizontal tail surface that pivots up and down about a central hinge point.

The canard design dates back to the pioneer days of aviation. Most notably, it was used on the Wright Flyer. Recently, the canard configuration has regained popularity and is appearing on newer aircraft. Canard designs include two types—one with a horizontal surface of about the same size as a normal aft-tail design, and the other with a surface of the same approximate size and airfoil of the aft-mounted wing known as a tandem wing configuration. Theoretically, the canard is considered more efficient because using the horizontal surface to help lift the weight of the aircraft should result in less drag for a given amount of lift.

The Piaggio P180 includes a variable-sweep canard design
that provides longitudinal stability about the lateral axis.

Rudder

The rudder controls movement of the aircraft about its vertical axis. This motion is called yaw. Like the other primary control surfaces, the rudder is a movable surface hinged to a fixed surface in this case, to the vertical stabilizer or fin. The rudder is controlled by the left and right rudder pedals.

When the rudder is deflected into the airflow, a horizontal force is exerted in the opposite direction. By pushing the left pedal, the rudder moves left. This alters the airflow around the vertical stabilizer/rudder and creates a sideward lift that moves the tail to the right and yaws the nose of the airplane to the left. Rudder effectiveness increases with speed; therefore, large deflections at low speeds and small deflections at high speeds may be required to provide the desired reaction. In propeller-driven aircraft, any slipstream flowing over the rudder increases its effectiveness.

V-Tail

The V-tail design utilizes two slanted tail surfaces to perform the same functions as the surfaces of a conventional elevator and rudder configuration. The fixed surfaces act as both horizontal and vertical stabilizers.

The movable surfaces, which are usually called ruddervators, are connected through a special linkage that allows the control wheel to move both surfaces simultaneously. On the other hand, displacement of the rudder pedals moves the surfaces differentially, thereby providing directional control.

When both rudder and elevator controls are moved by the pilot, a control mixing mechanism moves each surface the appropriate amount. The control system for the V-tail is more complex than the control system for a conventional tail. In addition, the V-tail design is more susceptible to Dutch roll tendencies than a conventional tail, and total reduction in drag is minimal.

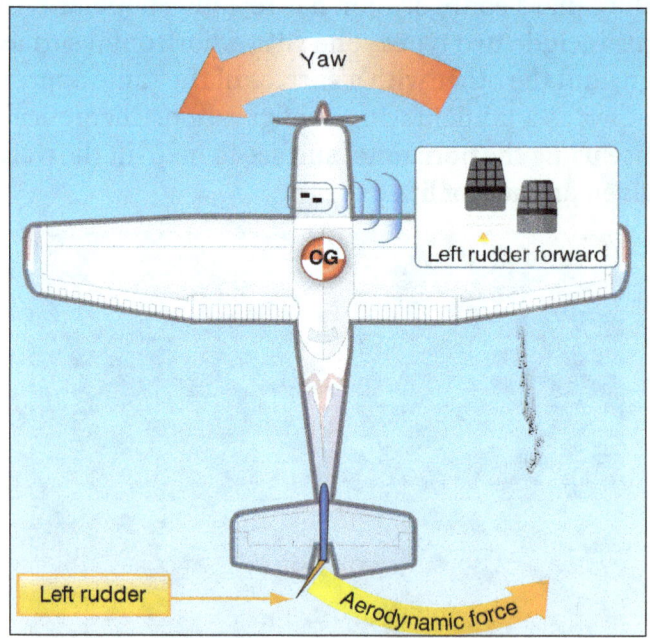

The effect of left rudder pressure.

Beechcraft Bonanza V35.

Secondary Flight Controls

Secondary flight control systems may consist of wing flaps, leading edge devices, spoilers, and trim systems.

Flaps

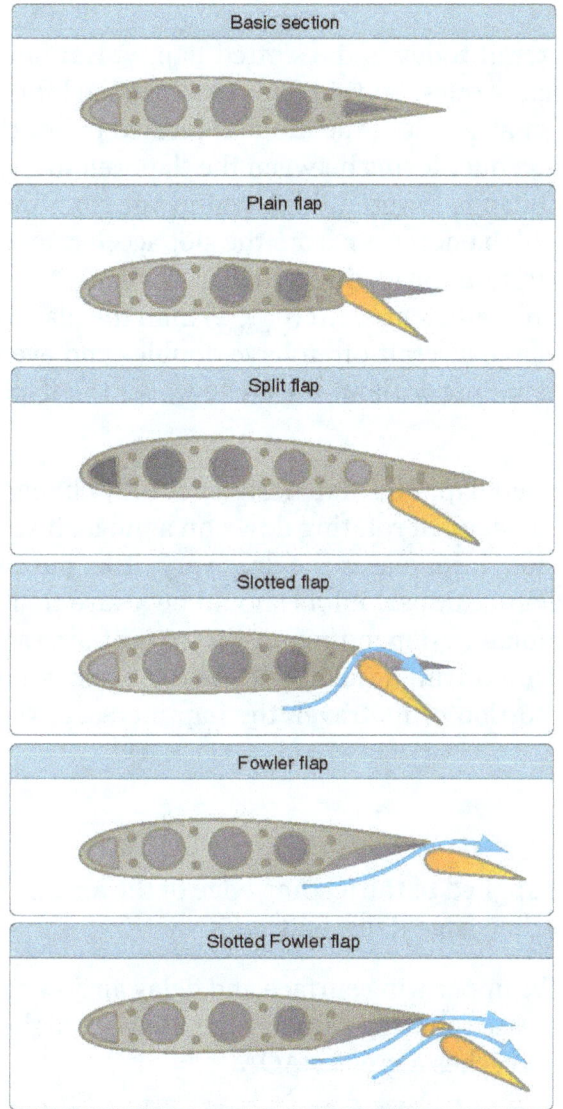

Five common types of flaps.

Flaps are the most common high-lift devices used on aircraft. These surfaces, which are attached to the trailing edge of the wing, increase both lift and induced drag for any given AOA. Flaps allow a compromise between high cruising speed and low landing speed because they may be extended when needed and retracted into the wing's structure when not needed. There are four common types of flaps: plain, split, slotted, and Fowler flaps.

The plain flap is the simplest of the four types. It increases the airfoil camber, resulting in a significant increase in the coefficient of lift (C_L) at a given AOA. At the same time, it greatly increases drag and moves the center of pressure (CP) aft on the airfoil, resulting in a nose-down pitching moment.

The split flap is deflected from the lower surface of the airfoil and produces a slightly greater increase in lift than the plain flap. More drag is created because of the turbulent air pattern produced behind the airfoil. When fully extended, both plain and split flaps produce high drag with little additional lift.

The most popular flap on aircraft today is the slotted flap. Variations of this design are used for small aircraft, as well as for large ones. Slotted flaps increase the lift coefficient significantly more than plain or split flaps. On small aircraft, the hinge is located below the lower surface of the flap, and when the flap is lowered, a duct forms between the flap well in the wing and the leading edge of the flap. When the slotted flap is lowered, high energy air from the lower surface is ducted to the flap's upper surface. The high energy air from the slot accelerates the upper surface boundary layer and delays airflow separation, providing a higher C_L. Thus, the slotted flap produces much greater increases in maximum coefficient of lift (C_{L-MAX}) than the plain or split flap. While there are many types of slotted flaps, large aircraft often have double- and even triple-slotted flaps. These allow the maximum increase in drag without the airflow over the flaps separating and destroying the lift they produce.

Fowler flaps are a type of slotted flap. This flap design not only changes the camber of the wing, it also increases the wing area. Instead of rotating down on a hinge, it slides backwards on tracks. In the first portion of its extension, it increases the drag very little, but increases the lift a great deal as it increases both the area and camber. Pilots should be aware that flap extension may cause a nose-up or down pitching moment, depending on the type of aircraft, which the pilot will need to compensate for, usually with a trim adjustment. As the extension continues, the flap deflects downward. During the last portion of its travel, the flap increases the drag with little additional increase in lift.

Leading Edge Devices

High-lift devices also can be applied to the leading edge of the airfoil. The most common types are fixed slots, movable slats, leading edge flaps, and cuffs.

Fixed slots direct airflow to the upper wing surface and delay airflow separation at higher angles of attack. The slot does not increase the wing camber, but allows a higher maximum C_L because the stall is delayed until the wing reaches a greater AOA.

Movable slats consist of leading edge segments that move on tracks. At low angles of attack, each slat is held flush against the wing's leading edge by the high pressure that forms at the wing's leading edge. As the AOA increases, the highpressure area moves aft below the lower surface of the wing, allowing the slats to move forward. Some slats, however, are pilot operated and can be deployed at any AOA. Opening a slat allows the air below the wing to flow over the wing's upper surface, delaying airflow separation.

Leading edge flaps, like trailing edge flaps, are used to increase both C_{L-MAX} and the camber of the wings. This type of leading edge device is frequently used in conjunction with trailing edge flaps and can reduce the nose-down pitching movement produced by the latter. As is true with trailing edge flaps, a small increment of leading edge flaps increases lift to a much greater extent than drag. As flaps are extended, drag increases at a greater rate than lift.

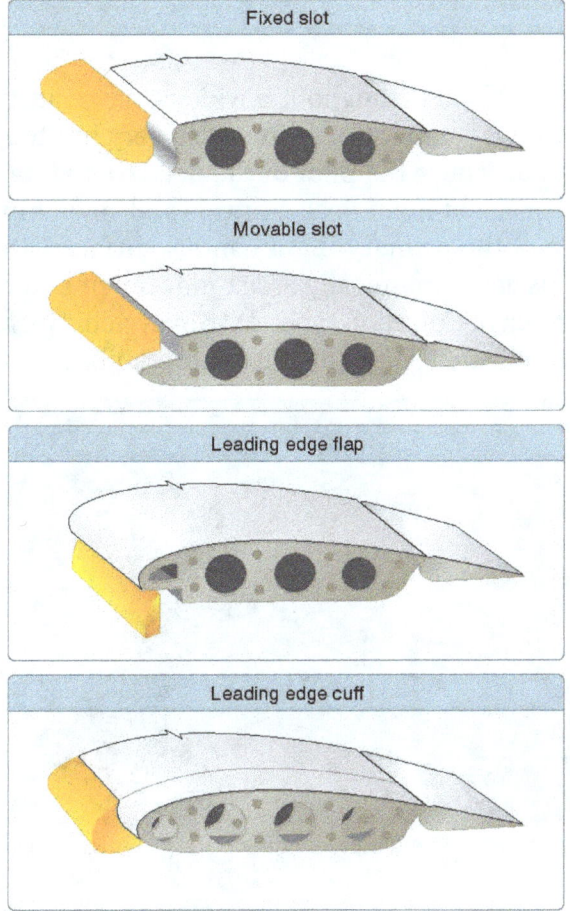

Leading edge high lift devices.

Leading edge cuffs, like leading edge flaps and trailing edge flaps are used to increase both $C_{L\text{-}MAX}$ and the camber of the wings. Unlike leading edge flaps and trailing edge flaps, leading edge cuffs are fixed aerodynamic devices. In most cases, leading edge cuffs extend the leading edge down and forward. This causes the airflow to attach better to the upper surface of the wing at higher angles of attack, thus lowering an aircraft's stall speed. The fixed nature of leading edge cuffs extracts a penalty in maximum cruise airspeed, but recent advances in design and technology have reduced this penalty.

Spoilers

Found on some fixed-wing aircraft, high drag devices called spoilers are deployed from the wings to spoil the smooth airflow, reducing lift and increasing drag. On gliders, spoilers are most often used to control rate of descent for accurate landings. On other aircraft, spoilers are often used for roll control, an advantage of which is the elimination of adverse yaw. To turn right, for example, the spoiler on the right wing is raised, destroying some of the lift and creating more drag on the right. The right wing drops, and the aircraft banks and yaws to the right. Deploying spoilers on both wings at the same time allows the aircraft to descend without gaining speed. Spoilers are also deployed to help reduce ground roll after landing. By destroying lift, they transfer weight to the wheels, improving braking effectiveness.

Trim Systems

Although an aircraft can be operated throughout a wide range of attitudes, airspeeds, and power settings, it can be designed to fly hands-off within only a very limited combination of these variables. Trim systems are used to relieve the pilot of the need to maintain constant pressure on the flight controls, and usually consist of flight deck controls and small hinged devices attached to the trailing edge of one or more of the primary flight control surfaces. Designed to help minimize a pilot's workload, trim systems aerodynamically assist movement and position of the flight control surface to which they are attached. Common types of trim systems include trim tabs, balance tabs, antiservo tabs, ground adjustable tabs, and an adjustable stabilizer.

Spoilers reduce lift and increase drag during descent and landing.

Trim Tabs

The most common installation on small aircraft is a single trim tab attached to the trailing edge of the elevator. Most trim tabs are manually operated by a small, vertically mounted control wheel. However, a trim crank may be found in some aircraft. The flight deck control includes a trim tab position indicator. Placing the trim control in the full nose-down position moves the trim tab to its full up position. With the trim tab up and into the airstream, the airflow over the horizontal tail surface tends to force the trailing edge of the elevator down. This causes the tail of the aircraft to move up and the nose to move down.

If the trim tab is set to the full nose-up position, the tab moves to its full down position. In this case, the air flowing under the horizontal tail surface hits the tab and forces the trailing edge of the elevator up, reducing the elevator's AOA. This causes the tail of the aircraft to move down and the nose to move up.

In spite of the opposing directional movement of the trim tab and the elevator, control of trim is natural to a pilot. If the pilot needs to exert constant back pressure on a control column, the need for nose-up trim is indicated. The normal trim procedure is to continue trimming until the aircraft is balanced and the nose-heavy condition is no longer apparent. Pilots normally establish the desired power, pitch attitude, and configuration first, and then trim the aircraft to relieve control pressures that may exist for that flight condition. As power, pitch attitude, or configuration changes, retrimming is necessary to relieve the control pressures for the new flight condition.

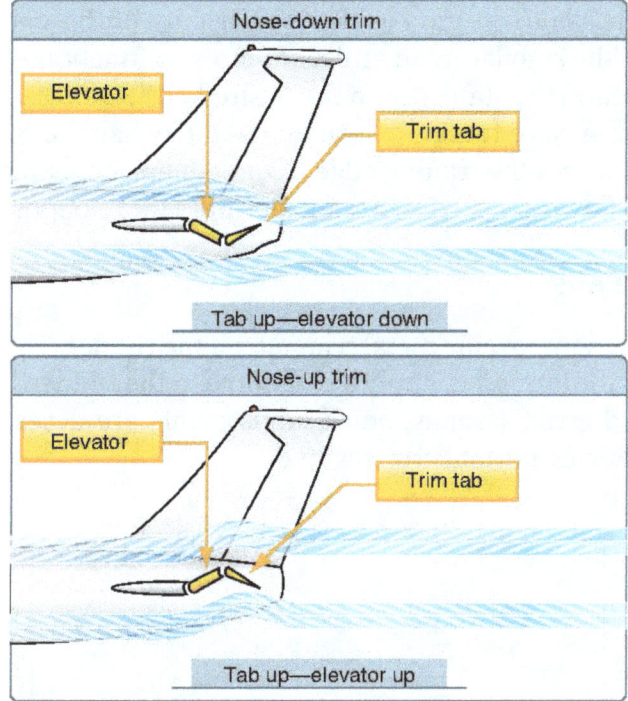

The movement of the elevator is opposite to the direction of movement of the elevator trim tab.

Balance Tabs

The control forces may be excessively high in some aircraft, and, in order to decrease them, the manufacturer may use balance tabs. They look like trim tabs and are hinged in approximately the same places as trim tabs. The essential difference between the two is that the balancing tab is coupled to the control surface rod so that when the primary control surface is moved in any direction, the tab automatically moves in the opposite direction. The airflow striking the tab counterbalances some of the air pressure against the primary control surface and enables the pilot to move the control more easily and hold the control surface in position.

If the linkage between the balance tab and the fixed surface is adjustable from the flight deck, the tab acts as a combination trim and balance tab that can be adjusted to a desired deflection.

Servo Tabs

Servo tabs are very similar in operation and appearance to the trim tabs previously discussed. A servo tab is a small portion of a flight control surface that deploys in such a way that it helps to move the entire flight control surface in the direction that the pilot wishes it to go. A servo tab is a dynamic device that deploys to decrease the pilots work load and de-stabilize the aircraft. Servo tabs are sometimes referred to as flight tabs and are used primarily on large aircraft. They aid the pilot in moving the control surface and in holding it in the desired position. Only the servo tab moves in response to movement of the pilot's flight control, and the force of the airflow on the servo tab then moves the primary control surface.

Antiservo Tabs

Antiservo tabs work in the same manner as balance tabs except, instead of moving in the opposite direction, they move in the same direction as the trailing edge of the stabilator. In addition to decreasing the sensitivity of the stabilator, an antiservo tab also functions as a trim device to relieve control pressure and maintain the stabilator in the desired position. The fixed end of the linkage is on the opposite side of the surface from the horn on the tab; when the trailing edge of the stabilator moves up, the linkage forces the trailing edge of the tab up. When the stabilator moves down, the tab also moves down. Conversely, trim tabs on elevators move opposite of the control surface.

Ground Adjustable Tabs

Many small aircraft have a nonmovable metal trim tab on the rudder. This tab is bent in one direction or the other while on the ground to apply a trim force to the rudder. The correct displacement is determined by trial and error. Usually, small adjustments are necessary until the aircraft no longer skids left or right during normal cruising flight.

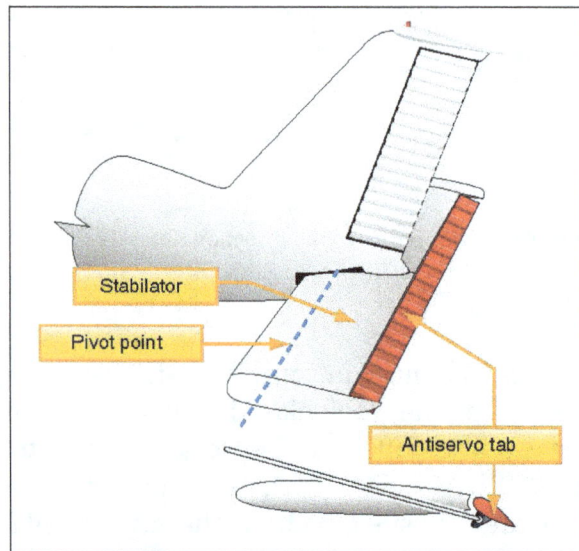

An antiservo tab attempts to streamline the control surface and is used to make the stabilator less sensitive by opposing the force exerted by the pilot.

Adjustable Stabilizer

Rather than using a movable tab on the trailing edge of the elevator, some aircraft have an adjustable stabilizer. With this arrangement, linkages pivot the horizontal stabilizer about its rear spar. This is accomplished by the use of a jackscrew mounted on the leading edge of the stabilator. On small aircraft, the jackscrew is cable operated with a trim wheel or crank. On larger aircraft, it is motor driven. The trimming effect and flight deck indications for an adjustable stabilizer are similar to those of a trim tab.

Autopilot

Autopilot is an automatic flight control system that keeps an aircraft in level flight or on a set course. It can be directed by the pilot, or it may be coupled to a radio navigation signal. Autopilot reduces the physical and mental demands on a pilot and increases safety. The common features available on an autopilot are altitude and heading hold.

The simplest systems use gyroscopic attitude indicators and magnetic compasses to control servos connected to the flight control system. The number and location of these servos depends on the complexity of the system. For example, a single-axis autopilot controls the aircraft about the longitudinal axis and a servo actuates the ailerons. A three-axis autopilot controls the aircraft about the longitudinal, lateral, and vertical axes. Three different servos actuate ailerons, elevator, and rudder. More advanced systems often include a vertical speed and indicated airspeed hold mode. Advanced autopilot systems are coupled to navigational aids through a flight director.

A ground adjustable tab is used on the rudder of many small airplanes to correct for a tendency to fly with the fuselage slightly misaligned with the relative wind.

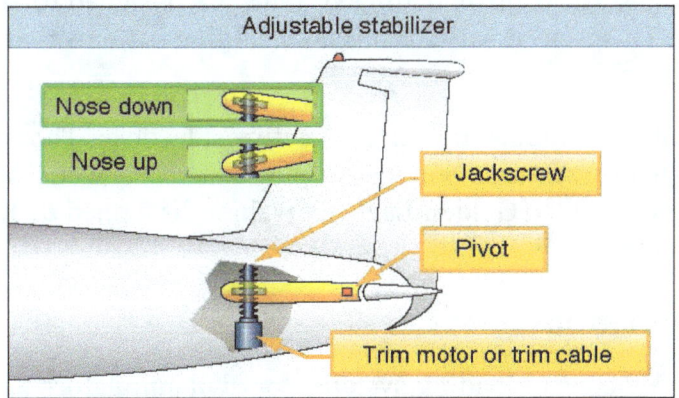

Some aircraft, including most jet transports, use an adjustable stabilizer to provide the required pitch trim forces.

The autopilot system also incorporates a disconnect safety feature to disengage the system automatically or manually. These autopilots work with inertial navigation systems, global positioning systems (GPS), and flight computers to control the aircraft. In fly-by-wire systems, the autopilot is an integrated component.

Additionally, autopilots can be manually overridden. Because autopilot systems differ widely in their operation, refer to the autopilot operating instructions in the Airplane Flight Manual (AFM) or the Pilot's Operating Handbook (POH).

CONTROL LOADING SYSTEM

A Control Loading System (CLS, also known as Electric Control Loading), is used to provide pilots with realistic flight control forces in a flight simulator or training device. These are used in both commercial and military training applications.

Design and Technology

The main concept is to provide forces to the pilot using an actuator (hydraulic or electric). The approach used in high fidelity applications is to connect this actuator via a linkage to the pilot controls. The actuator is then controlled with a servo controller to control the torque or current of the motor. An outer-loop control then controls the torque provided to the pilot using a control loop around a force sensor.

The control loading system must take in inputs from the simulator and pilot and provide outputs for the pilot and simulator. Inputs are application of force and aircraft states and outputs are flight control position and forces. An aircraft with reversible controls needs to have all of the complex components modeled within the control loading system. These include cables, rods, aero forces from the control surface, centering springs and trim actuators. As the control system gets more complicated they have to simulate effects such as bob-weights and feel units. Fly-by-wire systems are disconnected from the control surfaces and so do not need the complex features but add other functionality which is simulated. The high fidelity architecture has centralized control, individual analog signals to the control module, a brushless DC motor with low gear ratio and linkages to the pilot controls. The modular designs have localized control and digital reporting over a field bus to the central control module. The control loading systems are designed to allow situating the actuators closer to the pilot. This is necessary for mission training systems that can be easily deployed and moved around the world.

Control Loading Systems are similar in design to active sidesticks. These provide cues to pilots during the flight via actuation systems. Some examples of active sidesticks used in aircraft are for the F-35 Lightning II and the T-50 Golden Eagle jet trainer developed by KAI in partnership with Lockheed Martin Corporation.

Standards and Regulations

The regulations governing control loading systems for civil simulators are the Federal Aviation Administration regulations in North America and EASA (formerly JAA) in Europe. The FAA documents are AC 120-40B for airplane simulator qualification, Advisory circular 120-45A for Airplane Flight Training Device Qualification and AC 120-63 for helicopter Simulator Qualification. The EASA regulations are similar to the FAR's. Between 2006 and 2008 the International Working Group of the RAeS's Flight Simulation Group met on several occasions to redefine the standards applicable to flight simulation. This resulted in the release of a draft standards document to ICAO. This will be released by ICAO in 2009 and at this time the FAA and EASA should incorporate this into the regulations. The changes behind the standards will define different levels of simulator training devices which define what training requirements can be trained on with particular levels of simulators.

ELECTRONIC FLIGHT INSTRUMENT SYSTEM

An electronic flight instrument system (EFIS) is a flight deck instrument display system that displays flight data electronically rather than electromechanically. An EFIS normally consists of a primary flight display (PFD), multi-function display (MFD), and an engine indicating and crew alerting system (EICAS) display. Early EFIS models used cathode ray tube (CRT) displays, but liquid crystal displays (LCD) are now more common. The complex electromechanical attitude director indicator (ADI) and horizontal situation indicator (HSI) were the first candidates for replacement by EFIS. Now, however, few flight deck instruments cannot be replaced by an electronic display.

EFIS installations vary greatly. A light aircraft might be equipped with one display unit that displays flight and navigation data. A large, commercial aircraft is likely to have six or more display units. The equivalent electromechanical instruments are also shown here.

EFIS on an Airbus A380.

EFIS on an Eclipse 500.

EFIS installation follows the sequence:

- Displays,
- Controls,
- Data processors.

A basic EFIS might have all these facilities in the one unit.

Garmin G1000 on a Diamond DA42.

Primary Flight Display of a Boeing 747-400.

Display Units

Primary Flight Display (PFD)

On the flight deck, the display units are the most obvious parts of an EFIS system, and are the features that lead to the term *glass cockpit*. The display unit that replaces the ADI is called the primary flight display (PFD). If a separate display replaces the HSI, it is called the navigation display. The PFD displays all information critical to flight, including calibrated airspeed, altitude, heading, attitude, vertical speed and yaw. The PFD is designed to improve a pilot's situational awareness by integrating this information into a single display instead of six different analog instruments, reducing the amount of time necessary to monitor the instruments. PFDs also increase situational awareness by alerting the aircrew to unusual or potentially hazardous conditions — for example, low airspeed, high rate of descent — by changing the color or shape of the display or by providing audio alerts.

The names Electronic Attitude Director Indicator and Electronic Horizontal Situation Indicator are used by some manufacturers. However, a simulated ADI is only the centerpiece of the PFD. Additional information is both superimposed on and arranged around this graphic.

Multi-function displays can render a separate navigation display unnecessary. Another option is to use one large screen to show both the PFD and navigation display.

The PFD and navigation display (and multi-function display, where fitted) are often physically identical. The information displayed is determined by the system interfaces where the display units are fitted. Thus, spares holding is simplified: the one display unit can be fitted in any position.

LCD units generate less heat than CRTs; an advantage in a congested instrument panel. They are also lighter, and occupy a lower volume.

Multi-function Display (MFD)

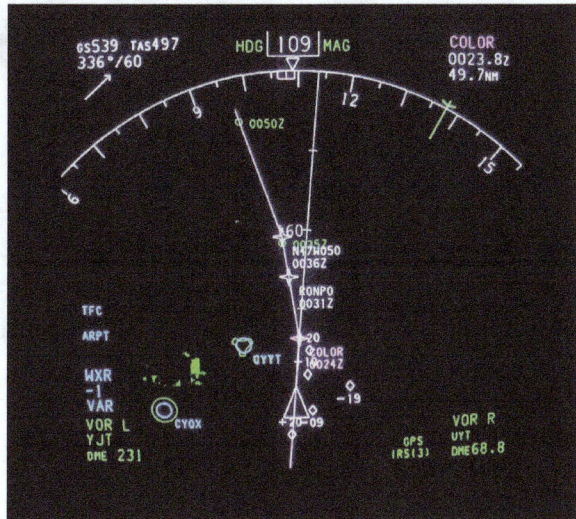

The Navigation Display (ND) of a Boeing 747-400 Aircraft.

The MFD (multi-function display) displays navigational and weather information from multiple systems. MFDs are most frequently designed as "chart-centric", where the aircrew can overlay different information over a map or chart. Examples of MFD overlay information include the aircraft's current route plan, weather information from either on-board radar or lightning detection sensors or ground-based sensors, e.g., NEXRAD, restricted airspace and aircraft traffic. The MFD can also be used to view other non-overlay type of data (e.g., current route plan) and calculated overlay-type data, e.g., the glide radius of the aircraft, given current location over terrain, winds, and aircraft speed and altitude.

MFDs can also display information about aircraft systems, such as fuel and electrical systems. As with the PFD, the MFD can change the color or shape of the data to alert the aircrew to hazardous situations.

Engine Indications and Crew Alerting System (EICAS) / Electronic Centralized Aircraft Monitoring (ECAM)

EICAS (Engine Indications and Crew Alerting System) displays information about the aircraft's systems, including its fuel, electrical and propulsion systems (engines). EICAS displays are often designed to mimic traditional round gauges while also supplying digital readouts of the parameters.

EICAS improves situational awareness by allowing the aircrew to view complex information in a graphical format and also by alerting the crew to unusual or hazardous situations. For example, if an engine begins to lose oil pressure, the EICAS might sound an alert, switch the display to the page with the oil system information and outline the low oil pressure data with a red box. Unlike traditional round gauges, many levels of warnings and alarms can be set. Proper care must be taken when designing EICAS to ensure that the aircrew are always provided with the most important information and not overloaded with warnings or alarms.

ECAM is a similar system used by Airbus, which in addition to providing EICAS functions also recommend remedial action.

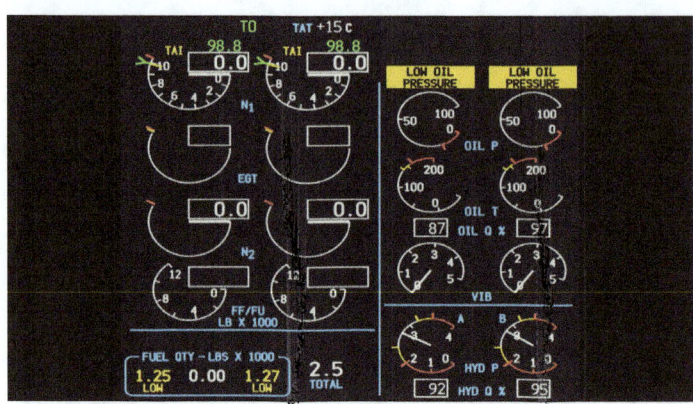

A 737NG EICAS after landing, showing outside air temperature, N1 RPM, exhaust gas temperature, N2 RPM, Fuel Flow/Fuel Used indications, fuel on the tanks, oil pressure, oil temperature, oil quantity, engine vibration, hydraulic pressure and hydraulic quantity.

Control Panels

EFIS provides pilots with controls that select display range and mode (for example, map or compass rose) and enter data (such as selected heading).

Where other equipment uses pilot inputs, data buses broadcast the pilot's selections so that the pilot need only enter the selection once. For example, the pilot selects the desired level-off altitude on a control unit. The EFIS repeats this selected altitude on the PFD, and by comparing it with the actual altitude (from the air data computer) generates an altitude error display. This same altitude selection is used by the automatic flight control system to level off, and by the altitude alerting system to provide appropriate warnings.

Data Processors

The EFIS visual display is produced by the symbol generator. This receives data inputs from the pilot, signals from sensors, and EFIS format selections made by the pilot. The symbol generator can go by other names, such as display processing computer, display electronics unit, etc.

The symbol generator does more than generate symbols. It has (at the least) monitoring facilities, a graphics generator and a display driver. Inputs from sensors and controls arrive via data buses, and are checked for validity. The required computations are performed, and the graphics generator and display driver produce the inputs to the display units.

Capabilities

Like personal computers, flight instrument systems need power-on-self-test facilities and continuous self-monitoring. Flight instrument systems, however, need additional monitoring capabilities:

- Input validation: Verify that each sensor is providing valid data.

- Data comparison: Cross check inputs from duplicated sensors.

- Display monitoring: Detect failures within the instrument system.

Former Practice

Traditional (electromechanical) displays are equipped with synchro mechanisms that transmit the pitch, roll, and heading shown on the captain and first officer's instruments to an instrument comparator. The comparator warns of excessive differences between the Captain and First Officer displays. Even a fault as far *downstream* as a jam in, say, the roll mechanism of an ADI triggers a comparator warning. The instrument comparator thus provides both comparator monitoring and display monitoring.

Comparator Monitoring

With EFIS, the comparator function is simple: Is roll data (bank angle) from sensor 1 the same as roll data from sensor 2? If not, display a warning caption (such as CHECK ROLL) on both PFDs. Comparison monitors give warnings for airspeed, pitch, roll, and altitude indications. More advanced EFIS systems have more comparator monitors.

Display Monitoring

In this technique, each symbol generator contains two display monitoring channels. One channel, the internal, samples the output from its own symbol generator to the display unit and computes, for example, what roll attitude should produce that indication. This computed roll attitude is then compared with the roll attitude input to the symbol generator from the INS or AHRS. Any difference has probably been introduced by faulty processing, and triggers a warning on the relevant display.

The external monitoring channel carries out the same check on the symbol generator on the other side of the flight deck: the Captain's symbol generator checks the First Officer's, the First Officer's checks the Captain's. Whichever symbol generator detects a fault, puts up a warning on its own display.

The external monitoring channel also checks sensor inputs (to the symbol generator) for reasonableness. A spurious input, such as a radio height greater than the radio altimeter's maximum, results in a warning.

Human Factors

Clutter

At various stages of a flight, a pilot needs different combinations of data. Ideally, the avionics only show the data in use—but an electromechanical instrument must be in view all the time. To improve display clarity, ADIs and HSIs use intricate mechanisms to remove superfluous indications temporarily—e.g., removing the glide slope scale when the pilot doesn't need it.

Under normal conditions, an EFIS might not display some indications, e.g., engine vibration. Only when some parameter exceeds its limits does the system display the reading. In similar fashion, EFIS is programmed to show the glideslope scale and pointer only during an ILS approach.

In the case of an input failure, an electromechanical instrument adds yet another indicator—typically, a bar drops across the erroneous data. EFIS, on the other hand, removes invalid data from the display and substitutes an appropriate warning.

A de-clutter mode activates automatically when circumstances require the pilot's attention for a specific item. For example, if the aircraft pitches up or down beyond a specified limit—usually 30 to 60 degrees—the attitude indicator de-clutters other items from sight until the pilot brings the pitch to an acceptable level. This helps the pilot focus on the most important tasks.

Color

Traditional instruments have long used color, but lack the ability to change a color to indicate some change in condition. The electronic display technology of EFIS has no such restriction and uses color widely. For example, as an aircraft approaches the glide slope, a blue caption can indicate glide slope is armed, and capture might change the color to green. Typical EFIS systems color code the navigation needles to reflect the type of navigation. Green needles indicate ground based navigation, such as VORs, Localizers and ILS systems. Magenta needles indicate GPS navigation.

Advantages

EFIS provides versatility by avoiding some physical limitations of traditional instruments. A pilot can switch the same display that shows a course deviation indicator to show the planned track provided by an area navigation or flight management system. Pilots can choose to superimpose the weather radar picture on the displayed route.

The flexibility afforded by software modifications minimises the costs of responding to new aircraft regulations and equipment. Software updates can update an EFIS system to extend its capabilities. Updates introduced in the 1990s included the ground proximity warning system and traffic collision avoidance system.

A degree of redundancy is available even with the simple two-screen EFIS installation. Should the PFD fail, transfer switching repositions its vital information to the screen normally occupied by the navigation display.

EMBEDDED INSTRUMENTATION

In the electronics industry, embedded instrumentation refers to the integration of test and measurement instrumentation into semiconductor chips (or integrated circuit devices). Embedded instrumentation differs from embedded system, which are electronic systems or subsystems that usually comprise the control portion of a larger electronic system. Instrumentation embedded into chips (embedded instrumentation) is employed in a variety of electronic test applications, including validating and testing chips themselves, validating, testing and debugging the circuit boards where these chips are deployed, and troubleshooting systems once they have been installed in the field.

Dating back to as early as the 1990s, the electronics industry recognized that design validation, test and debug would be seriously impeded in the near future. Initially, the impetus behind this recognition and the subsequent development of solutions was the emergence of new semiconductor chip packages such as the ball grid array (BGA) which placed the device's pins beneath the silicon die, making them inaccessible to physical contact with an instrument's or a test system's metal probes. At that time, most test instruments, such as the oscilloscope and logic analyzers in design, and in-circuit test (ICT) in volume manufacturing were external to the chips and circuit boards. They relied upon placing a probe on a chip or a circuit board to obtain test data. To overcome the disappearing access for test probes, instrumentation technology began to be embedded into semi-conductors and onto printed circuit boards.

In recent years, this situation has been exacerbated by increasingly high-speed serial inter-chip connections (interconnects or buses) on circuit boards as well as by even more complex chip packaging technologies like system-on-a-chip (SOC), system-in-package (SIP) and package-on-package (POP). These and other developments are making instrumentation embedded into chips a necessity for design validation, test and debug processes.

Boundary Scan Standard

Although it was not referred to as an embedded instrument at the time of its development, the IEEE 1149.1 Boundary Scan Standard can be seen as the first enabling technology for embedded instrumentation. (Boundary scan is also referred to as JTAG, after the Joint Test Action Group which initially undertook its development before it came under the aegis of a working group of the IEEE. 'JTAG' is often used to designate the access port on a chip which conforms to the boundary-scan standard.) Some would consider the boundary-scan test process as a form of embedded instrumentation. Boundary scan involves embedding an instrumentation infrastructure into chips and onto circuit boards. This infrastructure can be utilized during design debug and first prototype board bring-up, volume manufacturing and field service to test and diagnose the structural integrity of electrical connections on circuit boards. In addition, the boundary scan infrastructure can also be applied to the programming of devices such as memories, complex programmable logic devices (CPLDs) and Field-programmable gate arrays (FPGAs) after they have been soldered to a circuit board.

Related Standards

In the intervening years since the development of boundary-scan standard as a structural test technology, the embedded boundary-scan infrastructure in chips and on circuit boards has been appropriated for a number of related applications, including as an access method to the instrumentation inserted in chips. A later standard in the boundary-scan family, the IEEE 1149.6 Boundary-Scan Standard for Advanced Digital Networks, utilizes the 1149.1 boundary-scan embedded instrumentation infrastructure but expands the types of chip-to-chip interconnects that can be tested. Whereas the 1149.1 standard defines a methodology for testing DC-coupled interconnects, the 1149.6 version of the boundary-scan standard extends the methodology to testing high-speed AC coupled and differential interconnects.

Another addition to the boundary-scan family of standards has been IEEE 1149.7, which defines a reduced pin-count interface and provides for enhanced software debug. In addition, IEEE 1149.7 is expected to be used in the testing of complex chips with multiple die stacked in one package.

A working group of the IEEE has also undertaken the development of a standard that specifically addresses embedded instrumentation. The official name of this standard is the IEEE 1687 Standard for Access and Control of Instrumentation Embedded within a Semiconductor Device, but it is commonly referred to as the Internal JTAG (IJTAG) standard. The objective of the working group has been to define a technology for automating, accessing and analyzing the output of embedded instruments. Standardizing the interface to embedded instruments means that these instruments could come from any number of sources, but user interaction would be simplified since this would be based on an industry-accepted standard. The actual embedded instruments could be developed by any of several different types of groups, including chip suppliers, third-party providers, chip design tool vendors or in-house design groups. The availability of chips that conform to the IEEE 1687 standard would encourage the development of standards-based tools for interacting with and utilizing the instrumentation embedded in the chips. The IEEE 1687 standard is now a ratified standard.

The Emergence of Embedded Instrumentation

Embedded instrumentation can perform certain functions that external test and measurement technologies have difficulty with because they are external to the chip or circuit board being tested. Moreover, embedded instrumentation is often more efficient and adaptable, since it is software-based, unlike external hardware testers. In addition, embedded instrumentation is simply better suited to much of the computer and communications technologies that are emerging today.

Some of the more traditional test and measurement technologies are only able to measure performance and data flow at the input/output (I/O) point on a chip. The real challenge is for the instrument to gain visibility into the processing that is going on inside the core silicon itself. To do this, embedded instruments are needed between the I/O point on the perimeter of the chip and the processing core.

The following are some of the difficulties that traditional test and measurement instruments are running into as chips, circuit boards and systems continue to become faster, smaller and more complex.

- The signaling on high-speed serial buses like PCI Express, Fibre Channel 10-Gbit/s Ethernet, InfiniBand or Intel's QuickPath Interconnect (QPI) is very sensitive to capacitive coupling effects. As a result, placing a test pad on one of these buses to provide access to the test probes that external instruments such as oscilloscopes and in-circuit test systems rely on will disrupt the integrity of signals on the underlying bus. As a result, today's best practices for designing circuit boards call for the elimination of test pads, which restricts the use of test probes. The alternative is embedded instruments which are able to test from the inside out.

- Semiconductor vendors are incorporating signaling conditioning features such as pre-conditioning, pre-emphasis and equalization into their devices to help move signals at higher frequencies. Unfortunately, these techniques make it more difficult for traditional external instruments to take accurate measurements.

- Sub-100 nanometer chip fabrication processes have dramatic effects on device-level parametric performance characteristics to the extent that traditional characterization and testing techniques are often ineffective at identifying problems. External instruments at the corners of a chip cannot see the variations across the chip. On-chip or embedded instruments can effectively monitor parametric characteristics such as thermal conditions, timing issues, clock propagation delays, power distribution and others.

- External instruments typically only measure signal integrity margins on one or a few high-speed serial lanes at a time. Embedded instrumentation technologies such as Intel's IBIST (Interconnect Built-In Self Test) can test and measure all lanes on all buses concurrently. This makes the test more robust and more complete, and it reduces the amount of time required to validate the system.

- Embedded instrumentation can perform design validation, test and debug routines that external validation and test technologies cannot. An example of this would be Intel's IBIST technology which can stress and thereby test high-speed I/O buses well beyond the capabilities of traditional testing techniques that are applied through an operating system.

Embedded Instrumentation Applications

The applications for embedded instrumentation are extensive. At the level of circuit boards, two of the most prominent applications are design validation and non-intrusive board test.

Chip and Circuit Board Design Validation

Instruments are being embedded into chips and utilized to validate circuit board designs. This sort of design validation can make use of a variety of embedded instruments such as bit error rate test (BERT) engines, BIST) for logic devices, margining engines, memory BIST, memory test, random pattern generators and a complete logic analyzer. Deployed in design validation applications, these embedded instruments may function inside the chip or across on-board chip-to-chip interconnects to validate the performance and functionality of a circuit board design before it moves into volume production.

Non-Intrusive Board Test

Non-intrusive board test (NBT) employs embedded instrumentation to perform structural and electrical tests on circuit boards. In addition to boundary scan, other types of NBT methods include processor-controlled test (PCT) and FPGA-controlled test (FCT).

Embedded Instrumentation Methodologies

The IEEE 1149.1 Boundary-Scan Standard could be seen as the first enabler of embedded instrumentation and, as such, the first embedded instrumentation methodology. In addition to providing the infrastructure for accessing and operating embedded instruments, tests that utilize the boundary scan infrastructure can be applied to circuit boards to identify structural defects such as shorts and opens. Several other methodologies also apply tests that are initiated by embedded instruments.

- High-speed I/O (HSIO): In particular, Intel Corporation has developed embedded instrumentation technology which it is placing in all of its advanced processors. This can be employed to validate signal integrity on the high-speed buses in the Intel Architecture (IA).

- Processor-controlled test (PCT): PCT takes advantage of the debug port on most processor chips and asserts test and diagnostic routines on other elements on the circuit board, such as devices and the chip-to-chip buses.

- FPGA-controlled test (FCT): Instruments are temporarily embedded into a field program-mable gate array (FPGA) device on a circuit board to accomplish FCT tests. The types of tests performed will depend on the test functionality of the instruments embedded into the FPGA. Once the tests have been completed, the FCT embedded instruments can be removed and operating firmware loaded into the FPGA.

FIRE-CONTROL SYSTEM

A German anti-aircraft 88 mm Flak gun with its fire-control computer from World War II.
Displayed in the Canadian War Museum.

A fire-control system is a number of components working together, usually a gun data computer, a director, and radar, which is designed to assist a ranged weapon system in targeting, tracking and hitting its target. It performs the same task as a human gunner firing a weapon, but attempts to do so faster and more accurately.

Naval based Fire Control

The original fire-control systems were developed for ships.

The early history of naval fire control was dominated by the engagement of targets within visual range (also referred to as direct fire). In fact, most naval engagements before 1800 were con-ducted at ranges of 20 to 50 yards (20 to 50 m). Even during the American Civil War, the famous engagement between USS *Monitor* and CSS *Virginia* was often conducted at less than 100 yards (90 m) range.

Rapid technical improvements in the late 19th century greatly increased the range at which gun-fire was possible. Rifled guns of much larger size firing explosive shells of lighter relative weight (compared to all-metal balls) so greatly increased the range of the guns that the main problem became aiming them while the ship was moving on the waves. This problem was solved with the introduction of the gyroscope, which corrected this motion and provided sub-degree accuracies. Guns were now free to grow to any size, and quickly surpassed 10 inches calibre by the turn of the century. These guns were capable of such great range that the primary limitation was seeing the target, leading to the use of high masts on ships.

Another technical improvement was the introduction of the steam turbine which greatly increased the performance of the ships. Earlier screw-powered capital ships were capable of perhaps 16 knots, but the first large turbine ships were capable of over 20 knots. Combined with the long range of the guns, this meant that the ships moved a considerable distance, several ship lengths, between the time the shells were fired and landed. One could no longer *eyeball* the aim with any hope of accuracy. Moreover, in naval engagements it is also necessary to control the firing of several guns at once.

Naval gun fire control potentially involves three levels of complexity. Local control originated with primitive gun installations aimed by the individual gun crews. Director control aims all guns on the ship at a single target. Coordinated gunfire from a formation of ships at a single target was a focus of battleship fleet operations. Corrections are made for surface wind velocity, firing ship roll and pitch, powder magazine temperature, drift of rifled projectiles, individual gun bore diameter adjusted for shot-to-shot enlargement, and rate of change of range with additional modifications to the firing solution based upon the observation of preceding shots.

The resulting directions, known as a firing solution, would then be fed back out to the turrets for laying. If the rounds missed, an observer could work out how far they missed by and in which direction, and this information could be fed back into the computer along with any changes in the rest of the information and another shot attempted.

At first, the guns were aimed using the technique of artillery spotting. It involved firing a gun at the target, observing the projectile's point of impact (fall of shot), and correcting the aim based on where the shell was observed to land, which became more and more difficult as the range of the gun increased.

Between the American Civil War and 1905, numerous small improvements, such as telescopic sights and optical rangefinders, were made in fire control. There were also procedural improvements, like the use of plotting boards to manually predict the position of a ship during an engagement.

Then increasingly sophisticated mechanical calculators were employed for proper gun laying, typically with various spotters and distance measures being sent to a central plotting station deep within the ship. There the fire direction teams fed in the location, speed and direction of the ship and its target, as well as various adjustments for Coriolis effect, weather effects on the air, and other adjustments. Around 1905, mechanical fire control aids began to become available, such as the Dreyer table, Dumaresq (which was also part of the Dreyer table), and Argo Clock, but these devices took a number of years to become widely deployed. These devices were early forms of rangekeepers.

Arthur Pollen and Frederic Charles Dreyer independently developed the first such systems. Pollen began working on the problem after noting the poor accuracy of naval artillery at a gunnery practice near Malta in 1900. Lord Kelvin, widely regarded as Britain's leading scientist first proposed using an analogue computer to solve the equations which arise from the relative motion of the ships engaged in the battle and the time delay in the flight of the shell to calculate the required trajectory and therefore the direction and elevation of the guns.

Pollen aimed to produce a combined mechanical computer and automatic plot of ranges and rates for use in centralised fire control. To obtain accurate data of the target's position and relative motion, Pollen developed a plotting unit (or plotter) to capture this data. To this he added a gyroscope to allow for the yaw of the firing ship. Like the plotter, the primitive gyroscope of the time required substantial development to provide continuous and reliable guidance. Although the trials in 1905 and 1906 were unsuccessful, they showed promise.

Meanwhile, a group led by Dreyer designed a similar system. Although both systems were ordered for new and existing ships of the Royal Navy, the Dreyer system eventually found most favour with the Navy in its definitive Mark IV* form. The addition of director control facilitated a full, practicable fire control system for World War I ships, and most RN capital ships were so fitted by mid 1916. The director was high up over the ship where operators had a superior view over any gunlayer in the turrets. It was also able to co-ordinate the fire of the turrets so that their combined fire worked together. This improved aiming and larger optical rangefinders improved the estimate of the enemy's position at the time of firing. The system was eventually replaced by the improved "Admiralty Fire Control table" for ships built after 1927.

Admiralty Fire Control table in the transmitting station of HMS *Belfast*.

During their long service life, rangekeepers were updated often as technology advanced, and by World War II they were a critical part of an integrated fire control system. The incorporation of radar into the fire control system early in World War II provided ships the ability to conduct effective gunfire operations at long range in poor weather and at night. For U.S. Navy gun fire control systems.

The use of director-controlled firing, together with the fire control computer, removed the control of the gun laying from the individual turrets to a central position; although individual gun mounts and multi-gun turrets would retain a local control option for use when battle damage limited director information transfer (these would be simpler versions called "turret tables" in the Royal Navy). Guns could then be fired in planned salvos, with each gun giving a slightly different trajectory. Dispersion of shot caused by differences in individual guns, individual projectiles, powder ignition sequences, and transient distortion of ship structure was undesirably large at typical naval engagement ranges. Directors high on the superstructure had a better view of the enemy than a turret mounted sight, and the crew operating them were distant from the sound and shock of the guns. Gun directors were topmost, and the ends of their optical rangefinders protruded from their sides, giving them a distinctive appearance.

Unmeasured and uncontrollable ballistic factors, like high altitude temperature, humidity, barometric pressure, wind direction and velocity, required final adjustment through observation of the fall of shot. Visual range measurement (of both target and shell splashes) was difficult prior to availability of Radar. The British favoured coincident rangefinders while the Germans favored the stereoscopic type. The former were less able to range on an indistinct target but easier on the operator over a long period of use, the latter the reverse.

Ford Mk 1 Ballistic Computer: The name rangekeeper began to become inadequate to describe the increasingly complicated functions of rangekeeper. The Mk 1 Ballistic Computer was the first rangekeeper that was referred to as a computer. Note the three pistol grips in the foreground. Those fired the ship's guns.

Submarines were also equipped with fire control computers for the same reasons, but their problem was even more pronounced; in a typical "shot", the torpedo would take one to two minutes to reach its target. Calculating the proper "lead" given the relative motion of the two vessels was very difficult, and torpedo data computers were added to dramatically improve the speed of these calculations.

In a typical World War II British ship the fire control system connected the individual gun turrets to the director tower (where the sighting instruments were located) and the analogue computer in the heart of the ship. In the director tower, operators trained their telescopes on the target; one telescope measured elevation and the other bearing. Rangefinder telescopes on a separate mounting measured the distance to the target. These measurements were converted by the Fire Control table into the bearings and elevations for the guns to fire upon. In the turrets, the gunlayers adjusted the elevation of their guns to match an indicator for the elevation transmitted from the Fire Control table—a turret layer did the same for bearing. When the guns were on target they were centrally fired.

Even with as much mechanization of the process, it still required a large human element; the Transmitting Station (the room that housed the Dreyer table) for HMS *Hood*'s main guns housed 27 crew.

Directors were largely unprotected from enemy fire. It was difficult to put much weight of armour so high up on the ship, and even if the armour did stop a shot, the impact alone would likely knock the instruments out of alignment. Sufficient armour to protect from smaller shells and fragments from hits to other parts of the ship was the limit.

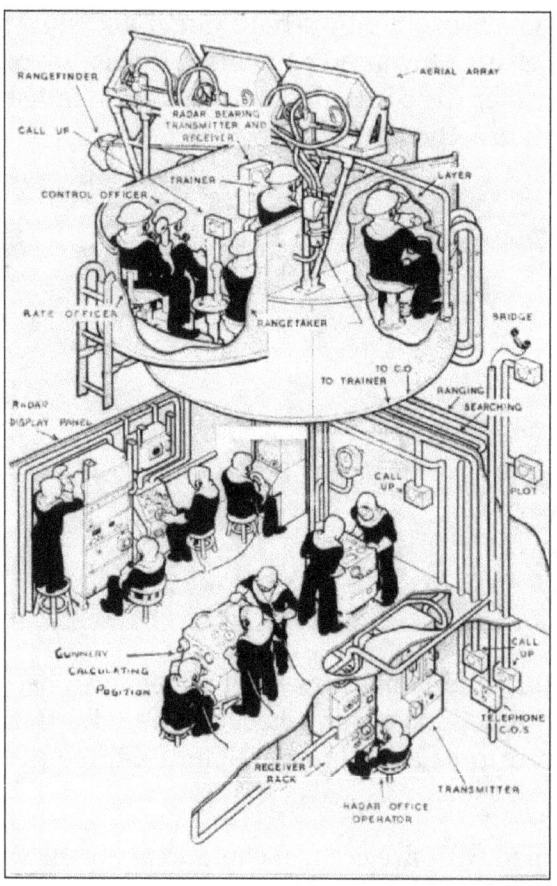

Accurate fire control systems were introduced in the early 20th century. Pictured, a cut-away view of a destroyer. The below deck analog computer is shown in the centre of the drawing and is labelled "Gunnery Calculating Position".

The performance of the analog computer was impressive. The battleship USS *North Carolina* during a 1945 test was able to maintain an accurate firing solution on a target during a series of high-speed turns. It is a major advantage for a warship to be able to maneuver while engaging a target.

Night naval engagements at long range became feasible when radar data could be input to the rangekeeper. The effectiveness of this combination was demonstrated in November 1942 at the Third Battle of Savo Island when the USS *Washington* engaged the Japanese battleship *Kirishima* at a range of 8,400 yards (7.7 km) at night. *Kirishima* was set aflame, suffered a number of explosions, and was scuttled by her crew. She had been hit by at least nine 16-inch (410 mm) rounds out of 75 fired (12% hit rate). The wreck of *Kirishima* was discovered in 1992 and showed that the entire bow section of the ship was missing. The Japanese during World War II did not develop radar or automated fire control to the level of the US Navy and were at a significant disadvantage.

By the 1950s gun turrets were increasingly unmanned, with gun laying controlled remotely from the ship's control centre using inputs from radar and other sources.

The last combat action for the analog rangekeepers, at least for the US Navy, was in the 1991 Persian Gulf War when the rangekeepers on the *Iowa*-class battleships directed their last rounds in combat.

Land based Fire Control

Anti-aircraft based Fire Control

By the start of World War II, aircraft altitude performance had increased so much that anti-aircraft guns had similar predictive problems, and were increasingly equipped with fire-control computers. The main difference between these systems and the ones on ships was size and speed. The early versions of the High Angle Control System, or HACS, of Britain's Royal Navy were examples of a system that predicted based upon the assumption that target speed, direction, and altitude would remain constant during the prediction cycle, which consisted of the time to fuze the shell and the time of flight of the shell to the target. The USN Mk 37 system made similar assumptions except that it could predict assuming a constant rate of altitude change. The Kerrison Predictor is an example of a system that was built to solve laying in "real time", simply by pointing the director at the target and then aiming the gun at a pointer it directed. It was also deliberately designed to be small and light, in order to allow it to be easily moved along with the guns it served.

The radar-based M-9/SCR-584 Anti-Aircraft System was used to direct air defense artillery since 1943. The MIT Radiation Lab's SCR-584 was the first radar system with automatic following, Bell Laboratory's M-9 was an electronic analog fire-control computer that replaced complicated and difficult-to-manufacture mechanical computers (such as the Sperry M-7 or British Kerrison predictor). In combination with the VT proximity fuze, this system accomplished the astonishing feat of shooting down V-1 cruise missiles with less than 100 shells per plane (thousands were typical in earlier AA systems). This system was instrumental in the defense of London and Antwerp against the V-1.

Coast Artillery Fire Control

In the United States Army Coast Artillery Corps, Coast Artillery fire control systems began to be developed at the end of the 19th Century and progressed on through World War II.

Early systems made use of multiple observation or base end stations to find and track targets attacking American harbors. Data from these stations were then passed to plotting rooms, where analog mechanical devices, such as the plotting board, were used to estimate targets' positions and derive firing data for batteries of coastal guns assigned to interdict them.

U.S. Coast Artillery forts bristled with a variety of armament, ranging from 12-inch coast defense mortars, through 3-inch and 6-inch mid-range artillery, to the larger guns, which included 10-inch and 12-inch barbette and disappearing carriage guns, 14-inch railroad artillery, and 16-inch cannon installed just prior to and up through World War II.

Fire control in the Coast Artillery became more and more sophisticated in terms of correcting firing data for such factors as weather conditions, the condition of powder used, or the Earth's rotation. Provisions were also made for adjusting firing data for the observed fall of shells. As shown in figure, all of these data were fed back to the plotting rooms on a finely tuned schedule controlled by a system of time interval bells that rang throughout each harbor defense system.

It was only later in World War II that electro-mechanical gun data computers, connected to coast defense radars, began to replace optical observation and manual plotting methods in controlling coast artillery. Even then, the manual methods were retained as a back-up through the end of the war.

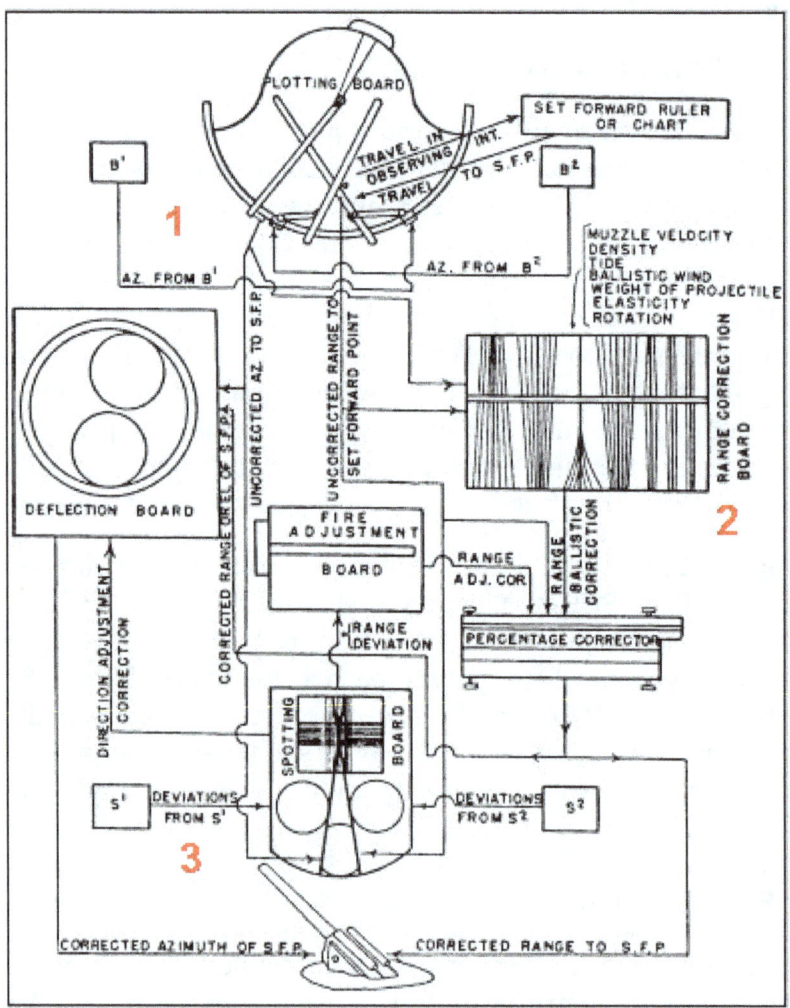

A conceptual diagram of the flow of fire control data in the Coast Artillery. The set forward point of the target was generated by using the plotting board (1). This position was then corrected for factors affecting range and azimuth (2). Finally, fire was adjusted for observations of the actual fall of the shells (3), and new firing data were sent to the guns.

Direct and Indirect Fire Control Systems

Land based fire control systems can be used to aid in both Direct fire and Indirect fire weapon engagement. These systems can be found on weapons ranging from small handguns to large artillery weapons.

Modern Fire Control Systems

Modern fire-control computers, like all high-performance computers, are digital. The added performance allows basically any input to be added, from air density and wind, to wear on the barrels and distortion due to heating. These sorts of effects are noticeable for any sort of gun, and fire-control computers have started appearing on smaller and smaller platforms. Tanks were one early use that automated gun laying using a laser rangefinder and a barrel-distortion meter. Fire-control computers are not just useful for large cannons. They can be used to aim machine guns, small cannons, guided missiles, rifles, grenades, rockets—any kind of weapon that can have its launch or firing parameters varied. They are typically installed on ships, submarines, aircraft, tanks and even on some small arms—for example, the grenade launcher developed for use on the Fabrique Nationale F2000 bullpup assault rifle. Fire-control computers have gone through all the stages of technology that computers have, with some designs based upon analogue technology and later vacuum tubes which were later replaced with transistors.

Fire-control systems are often interfaced with sensors (such as sonar, radar, infra-red search and track, laser range-finders, anemometers, wind vanes, thermometers, barometers, etc.) in order to cut down or eliminate the amount of information that must be manually entered in order to calculate an effective solution. Sonar, radar, IRST and range-finders can give the system the direction to and distance of the target. Alternatively, an optical sight can be provided that an operator can simply point at the target, which is easier than having someone input the range using other methods and gives the target less warning that it is being tracked. Typically, weapons fired over long ranges need environmental information—the farther a munition travels, the more the wind, temperature, air density, etc. will affect its trajectory, so having accurate information is essential for a good solution. Sometimes, for very long-range rockets, environmental data has to be obtained at high altitudes or in between the launching point and the target. Often, satellites or balloons are used to gather this information.

Once the firing solution is calculated, many modern fire-control systems are also able to aim and fire the weapons. Once again, this is in the interest of speed and accuracy, and in the case of a vehicle like an aircraft or tank, in order to allow the pilot/gunner/etc. to perform other actions simultaneously, such as tracking the target or flying the aircraft. Even if the system is unable to aim the weapon itself, for example the fixed cannon on an aircraft, it is able to give the operator cues on how to aim. Typically, the cannon points straight ahead and the pilot must maneuver the aircraft so that it oriented correctly before firing. In most aircraft the aiming cue takes the form of a "pipper" which is projected on the heads-up display (HUD). The pipper shows the pilot where the target must be relative to the aircraft in order to hit it. Once the pilot maneuvers the aircraft so that the target and pipper are superimposed, he or she fires the weapon, or on some aircraft the weapon will fire automatically at this point, in order to overcome the delay of the pilot. In the case of a missile launch, the fire-control computer may give the pilot feedback about whether the target is in range of the missile and how likely the missile is to hit if launched at any particular moment. The pilot will then wait until the probability reading is satisfactorily high before launching the weapon.

INTELLIGENT FLIGHT CONTROL SYSTEM

The Intelligent Flight Control System (IFCS) is a next-generation flight control system designed to provide increased safety for the crew and passengers of aircraft as well as to optimize the aircraft performance under normal conditions. The main benefit of this system is that it will allow a pilot to control an aircraft even under failure conditions that would normally cause it to crash.

NASA's NF-15B was used for the project.

Objectives of IFCS

The main purpose of the IFCS project is to create a system for use in civilian and military aircraft that is both adaptive and fault tolerant. This is accomplished through the use of upgrades to the flight control software that incorporate self-learning neural network technology. The goals of the IFCS neural network project are:

- To develop a flight control system that can identify aircraft characteristics through the use of neural network technology in order to optimize aircraft performance.

- To develop a neural network that can train itself to analyze the flight properties of the aircraft.

- To be able to demonstrate the aforementioned properties on a modified F-15 ACTIVE aircraft during flight, which is the testbed for the IFCS project.

Theory of Operation

The neural network of the IFCS learns flight characteristics in real time through the aircraft's sensors and from error corrections from the primary flight computer, and then uses this information to create different flight characteristic models for the aircraft. The neural network only learns when the aircraft is in a stable flight condition, and will discard any characteristics that would cause the aircraft to go into a failure condition. If the aircraft's condition changes from stable to failure, for example, if one of the control surfaces becomes damaged and unresponsive, the IFCS can detect this fault and switch the flight characteristic model for the aircraft. The neural network then works to drive the error between the reference model and the actual aircraft state to zero.

Project History

Generation 1

Generation 1 IFCS flight tests were conducted in 2003 to test the outputs of the neural network. In this phase, the neural network was pre-trained using flight characteristics obtained for the McDonnell Douglas F-15 STOL/MTD in a wind tunnel test and did not actually provide any control adjustments in flight. The outputs of the neural network were run directly to instrumentation for data collection purposes only.

Generation 2

Generation 2 IFCS tests were conducted in 2005 and used a fully integrated neural network as described in the theory of operation. It is a direct adaptive system that continuously provides error corrections and then measures the effects of these corrections in order to learn new flight models or adjust existing ones. To measure the aircraft state, the neural network takes 31 inputs from the roll, pitch, and yaw axes and the control surfaces. If there is a difference between the aircraft state and model, the neural network adjusts the outputs of the primary flight computer through a dynamic inversion controller to bring the difference to zero before they are sent to the actuator control electronics which move the control surfaces.

GUIDANCE SYSTEM

Purpose and Function

Every missile guidance system consists of an attitude control system and a flight path control system. The attitude control system functions to maintain the missile in the desired attitude on the ordered flight path by controlling the missile in pitch, roll, and yaw. The attitude control system operates as an auto-pilot, damping out fluctuations that tend to deflect the missile from its ordered flight path. The function of the flight path control system is to determine the flight path necessary for target interception and to generate the orders to the attitude control system to maintain that path.

It should be clear at this point that the concept of "Guidance and Control" involves not only the maintenance of a particular vehicle's path from point A to B in space, but also the proper behavior of the vehicle while following the path. A missile that follows a prescribed path half the way to a target and then becomes dynamically unstable is then incapable of remaining upon the path (or else fails structurally due to aerodynamic loading). Such a vehicle, in order to perform properly, must be "piloted" and capable of responding to control signals.

The operation of a guidance and control system is based on the principle of feedback. The control units make corrective adjustments of the missile control surfaces when a guidance error is present.

The control units will also adjust the control surfaces to stabilize the missile in roll, pitch, and yaw. Guidance and stabilization corrections are combined, and the result is applied as an error signal to the control system.

Sensors

The guidance system in a missile can be compared to the human pilot of an airplane. As a pilot guides his plane to the landing field, the guidance system "sees" its target. If the target is far away or otherwise obscured, radio or radar beams can be used to locate it and direct the missile to it. Heat, light, television, the earth's magnetic field, and Loran have all been found suitable for specific guidance purposes. When an electromagnetic source is used to guide the missile, an antenna and a receiver are installed in the missile to form what is known as a sensor. The sensor picks up, or senses, the guidance information. Missiles that are guided by other than electromagnetic means use other types of sensors, but each must have some means of receiving "position reports."

The kind of sensor that is used will be determined by such factors as maximum operation range, operating conditions, the kind of information needed, the accuracy required, viewing angle, and weight and size of the sensor, and the type of target and its speed.

Accelerometers

The heart of the inertial navigation system for ships and missiles is an arrangement of accelerometers that will detect any change in vehicular motion. To understand the use of accelerometers in inertial guidance, it is helpful to examine the general principles involved.

An accelerometer, as its name implies, is a device for measuring acceleration. In their basic form such devices are simple. For example, a pendulum, free to swing on a transverse axis, could be used to measure acceleration along the fore-and-aft axis of the missile. When the missile is given a forward acceleration, the pendulum will tend to lag aft; the actual dis-placement of the pendulum form its original position will be a function of the magnitude of the accelerating force. Another simple device might consist of a weight supported between two springs. When an accelerating force is applied, the weight will move from its original position in a direction opposite to that of the applied force. The movement of the mass (weight) is in accordance with Newton's second law of motion, which states that the acceleration of a body is directly proportional to the force applied and inversely proportional to the mass of the body.

If the acceleration along the fore-and aft axis were constant, the speed of the missile at any instant could be deter-mined simply by multiplying the acceleration by the elapsed time. However, the acceleration may change considerably over a period of time. Under these conditions, integration is necessary to determine the speed.

If the missile speed were constant, the distance covered could be calculated simply by multiplying the speed by time of flight. But because the acceleration varies, the speed also varies. For that reason, a second integration is necessary.

The moving element of the accelerometer can connected to a potentiometer, or to a variable inductor core, or to some other device capable of producing a voltage proportional to the dis-placement of the element.

Usually there are three double-integrating accelerometers continuously measuring the distance traveled by the missile in three directions--range, altitude, and azimuth. Double-integrating accelerometers are devices that are sensitive to acceleration, and by a double-step process measure distance. These measured distances are then compared with the desired distances, which are preset into the missile; if the missile is off course, correction signals are sent to the control system.

Accelerometers are sensitive to the acceleration of gravity as well as missile accelerations. For this reason, the accelerometers that measure range and azimuth distances must be mounted in a fixed position with respect to the pull of gravity. This can be done in a moving missile by mounting them on a platform that is stabilized by gyroscopes or by star-tracking telescopes. This platform, however, must be moved as the missile passes over the earth to keep the sensitive axis of each accelerometer in a fixed position with respect to the pull of gravity. These factors cause the accuracy of the inertial system to decrease as the time of flight of the missile increases.

To eliminate unwanted oscillations, a damper is included in the accelerometer unit. The damping effort should be just great enough to prevent any oscillations from occurring, but still permit a significant displacement of the mass. When this condition exists, the movement of the mass will be exactly proportional to the acceleration of the vehicle.

Figure shows a mass suspended by a spring in a liquid-damped system. If the case experiences an acceleration in the direction indicated by the arrow, the spring will offer a re-straining force proportional to the downward displacement of the mass, while the viscous fluid will serve to dampen any undesirable oscillations.

Figure shows a system that is electrically damped. The mass (M) is free to slide back and forth in relation to the iron core (C). When the vehicle experiences an acceleration, the voltage (E), which is proportional to the displacement of the mass, is picked off and amplified. The current (I) (still proportional to the displacement) is sent back to the coil around the core. The resulting magnetic field around the coil creates a force on the mass, which damps the oscillations. In this system the acceleration could be measured by the displacement of the mass (X), by the voltage (E), or by the current (I).

Phases of Guidance

Missile guidance is generally divided into three phases--boost, midcourse, and terminal. These names refer to different parts of the flight path. The boost phase may also be called the launching or initial phase.

Boost Phase

Navy surface-to-air missiles accelerate to flight speed by means of the booster component. This booster period lasts from the time the missile leaves the launcher until the booster burns its fuel. In missiles with separate boosters, the booster drops away from the missile at burnout. The objective of this phase is to place the missile at a position in space from where it can either "see" the target or where it can receive external guidance signals. During the boost phase of some missiles, the guidance system and the aerodynamic surfaces are locked in position. Other missiles are guided during the boost phase.

Midcourse Phase

The second, or midcourse, phase of guidance is often the longest in both distance and time. During this part of the flight, changes may be required to bring the missile onto the desired course and to make certain that it stays on that course. During this guidance phase, information can be supplied to the missile by any of several means. In most cases, the midcourse guidance system is used to place the missile near the target, where the System to be used in the final phase of guidance can take over. In other cases, the midcourse guidance system is used for both the second and third guidance phases.

Terminal Phase

The last phase of missile guidance must have high accuracy as well as fast response to guidance signals. Missile performance becomes a critical factor during this phase. The missile must be capable of executing the final maneuvers required for intercept within the constantly decreasing available flight time. The maneuverability of the missile will be a function of velocity as well as airframe design. Therefore, a terminal guidance system must be compatible with missile performance capabilities. The greater the target acceleration, the more critical the method of terminal guidance becomes. In some missiles, especially short-range missiles, a single guidance system may be used for all three phases of guidance, whereas other missiles may have a different guidance system for each phase.

Types of Guidance Systems

Missile guidance systems may be classified into two broad categories: missiles guided by man-made electromagnetic devices, and those guided by other means. In the first category are those missiles controlled by radar, radio devices, and those missiles that use the target as a source of electromagnetic radiation. In the latter category are missiles that rely on electromechanical devices or electromagnetic contact with natural sources, such as the stars (self-contained guidance systems).

All of the missiles that maintain electromagnetic radiation contact with man-make sources may be further subdivided into two subcategories.

- Control guidance missiles
- Homing guidance missiles

Control Guidance

Control guidance missiles are those that are guided on the basis of direct electromagnetic radiation contact with friendly control points. Homing guidance missiles are those that guided on the basis of direct electromagnetic radiation contact with the tar-get. Control guidance generally depends on the use of radar (radar control) or radio (radio control) links between a control point and the missile. By use of guidance information transmitted from the control point via a radio or radar link, the missile's flight path can be guided.

Radar Control Guidance: Radar control guidance may be subdivided into two separate categories. The first category is simply referred to as the command guidance method. The second is the beam-rider method, which is actually a modification of the first, but with the radar being used in a different manner.

Command guidance: The term command is used to describe a guidance method in which all guidance instructions, or commands, come from sources outside the missile. The guidance system of the missile contains a receiver that is capable of receiving instructions from ship or ground stations or from aircraft. The missile flight-path control system then converts these commands to guidance information, which is fed to the attitude control system.

In the command guidance method, one or two radars are used to track the missile and target. Figure is a block diagram of how this method works in actual practice. As soon as the radar is locked on the target, tracking information is fed into the computer. The missile is then launched and is tracked by the radar. Target and missile ranges, elevations, and bearings are continuously fed to the computer.

This information is analyzed and a missile intercept flight path is computed. The appropriate guidance signals are then transmitted to the missile receiver. These signals may be sent by varying the characteristics of the missile-tracking radar beam, or by way of a separate radio transmitter. The radar command guidance method can be used in ship, air, or ground missile delivery systems. A relatively new type of command guidance by wire is now operational in some short-range anti-tank-type weapons. These systems use an optical sight for tracking the target while the weapons emits a characteristic infrared signature, which is used for tracking the weapon with an IR sensor. Deviation of the weapon from the line of sight (LOS) to the target is sensed, and guidance commands are generated that are fed to the weapon control system in flight via a direct wire link. Each weapon contains wire spools that pay out as the warhead flies out the line of sight to the target. Current usage of these systems is in relatively lightweight, portable, short-range battlefield environments against armored targets where their high accuracy and substantial warheads are most efficiently employed.

Beam-rider Method: The main difference between the beam-rider method and the radar command guidance method is that the characteristics of the missile-tracking radar beam are not varied in the beam-rider system. The missile has been designed so that it is able to formulate its own correction signals on the basis of its position with respect to the radar scan axis. The missile's flight path control unit is sensitive to any deviation from the scan axis of the guidance radar and is capable of computing the proper flight path correction. An advantage of this type of system is that is requires only one radar. This radar must, of course, have a conical-scan feature in order to provide both target-tracking capability and a missile flight-path correction reference axis. A second advantage is that since the missile formulates its own directional commands, several missiles may be launched to "ride" the beam simultaneously, without the need for a cumbersome and complicated multiple-missile command system.

The accuracy of this system decreases with range because the radar beam spreads out, and it is more difficult for the missile to remain in its center. If the target is moving very rapidly, the missile must follow a continuously changing path, which may cause it to undergo excessive transverse accelerations.

Homing Guidance

Homing guidance systems control the flight path by employing a device in the weapon that reacts to some distinguishing feature of the target. Homing devices can be made sensitive to a variety of energy forms, including RF, infrared, reflected laser, sound, and visible light. In order to home on the target, the missile or torpedo must determine at least the azimuth and elevation of the target by one of the means of angle tracking mentioned previously. Active homing missiles will also have the means of determining range of the target if necessary. Tracking is performed by a movable seeker antenna or an array with stationary electronically scanned arrays in development for missiles and operational in some torpedoes. Determination of angular error by amplitude comparison monopulse methods is preferred over the older COSRO systems because of the higher data rate and faster response time; however, phase comparison monopulse or interferometer methods have advantages in some applications. Homing guidance methods may be divided into three types: active, semiactive, and passive homing. These methods may be employed in seekers using any of the energy forms mentioned above, although some methods may be excluded by the nature of the energy form; for example, one would not build a passive laser seeker or an active or semi-active infrared seeker.

Active homing: In active homing, the weapon contains both the transmitter and receiver. Search and acquisition are conducted as with any tracking sensor. The target is tracked employing monostatic geometry in which the returning echo from the target travels the same path as the transmitted energy. An onboard computer calculates a course to intercept the target and sends steering commands to the weapon's autopilot. The monostatic geometry allows the most efficient reflection of energy from the target, but the small size of the missile restricts the designer to high frequencies and low power output from the transmitter, resulting in short seeker acquisition range.

Semiactive homing: In semiactive homing, the target is illuminated by a tracking radar at the launching site or other control point. The missile is equipped with a radar receiver (no transmitter) and by means of the reflected radar energy from the target, formulates its own correction signals as in the active method. However, semiactive homing uses bistatic reflection from the target, meaning that because the illuminator platform and weapon receiver are not co-located, the returning echo follows a different path than the energy incident to the target. Due to its shape and composition, the target may not reflect energy efficiently in the direction of the weapon. In extreme cases the weapon may lose the target entirely, resulting in a missed intercept. This disadvantage is compensated for by the ability to use greater power and more diverse frequency ranges in an illum- ination device in a ship, aircraft, or ground station.

Passive homing: Passive homing depends only on the target as a source of tracking energy. This energy can be the noise radiated by a ship or submarine in the case of a passive homing torpedo, RF radiation from the target's own sensors in the case of an anti-radiation (ARM) weapon, heat sources such as ship, aircraft, or vehicle exhausts, contrast with the temperature or visible light environment, or even the radiation all objects emit in the microwave region. As in the other homing methods, the missile generates its own correction signals on the basis of energy received from the target rather than from a control point. The advantage of passive homing is that the counter detection problem is reduced, and a wide range of energy forms and frequencies are available. Its disadvantages are its susceptibility to decoy or deception and its dependence on a certain amount of cooperation from the enemy.

Retransmission Homing or Track Via Missile (TVM). Re-transmission homing is a blending of the characteristics of both command and semiactive homing guidance. In command guidance, missile steering commands are computed at the launch point using target position and missile position data derived from launch point sensors. In retransmission homing, the missile contains a semiactive seeker that determines the azimuth and elevation angle from the missile to the target, which is then coded and transmitted to the launch point via data link (down link). The fire control system at the launch point can use its own target tracking data, that of the missile (or both), and missile position data to compute steering commands, which are then transmitted to the missile via an uplink. This technique is used in some new AAW missile systems, including the U.S. Army Patriot system. Specific retransmission or TVM systems may vary somewhat from this ideal; however, they all will in some way use target angle data from the missile to compute steering commands at the launch point that are then transmitted to the missile.

Accuracy: Homing is the most accurate of all guidance systems because it uses the target as its source when used against moving targets. There are several ways in which the homing device may control the path of a missile against a moving target. Of these, the more generally used are pursuit paths and lead flight paths. Because monopulse methods in weapons seekers are advantageous and are becoming the method of choice in current weapons, it is necessary to address the two basic types:

Amplitude Comparison Monopulse. This method, requires a gimballed seeker antenna covered by a radome at the nose of the weapon. Because of aerodynamic requirements, the radome shape is normally not optimal for radar performance. Very precise orders to the antenna are required to achieve target acquisition due to the single antenna's limited field of view. In these systems the size of the antenna directly determines the limits of the frequency range of the seeker. Its primary advantage is its consistent performance throughout the potential speed and maneuverability range of potential targets.

Interferometer (Phase Comparison Monopulse). The interferometer eliminates the requirement for a movable antenna, having instead fixed antennas mounted at the edge of the airframe or on the wing tips, the result being reduced complexity and a wider field of view. As depicted in figure, two antennas separated by a known distance are installed for each mobility axis of the weapon. In the diagram the antennas A and B, separated by the distance d, receive energy emitted (passive homing) or reflected (semiactive homing) from the target.

Because the distance to the target is relatively large, it is assumed that the RF energy arrives as a series of planar waves with wavelength . In accordance with the discussion of electronic scanning it is evident that for the geometry pictured, the phase sensed by antenna B will lag that sensed by antenna A by some phase angle which is proportional to d sin ; therefore:

$$\underline{2d} \sin$$

If is known and the phase angle can be determined, then the look angle, , can be calculated.

The interferometer provides the advantage of wide field of view, flexibility in airframe design, unobstructed use of weapon interior space, and the ability to cover broad frequency bands without constraints imposed by limited antenna size. The separation between the antennas governs the performance of the system, with missile body diameter or fin spread separation as the usual arrangement. The disadvantage of the interferometer is the angular ambiguity that may exist for wavelengths less than the separation between the antennas at a specific angle of incidence. If the distance between the antennas at an angle of incidence is d sin , and is less than d sin , then it is not possible to determine if the phase angle measured is just that or + n2 radians, where n is any integer. However, this is a minor problem in most homing systems because the absolute look angle is not as important as the rate of change of that angle.

The interferometer has an advantage in resolving multiple targets at twice the range of a typical amplitude comparison monopulse seeker in the same size weapon. This gives the missile twice the time to respond to the changeover from tracking the centroid of the group to tracking one specific target, thus in-creasing the hit probability.

Composite Systems. No one system is best suited for all phases of guidance. It is logical then to combine a system that has good midcourse guidance characteristics with a system that has excellent terminal guidance characteristics, in order to in-crease the number of hits. Combined systems are known as composite guidance systems or combination systems.

Many missiles rely on combinations of various types of guidance. For example, one type of missile may use command guidance until it is within a certain range of a target. At this time the command guidance may become a back-up mode and a type of homing guidance commenced. The homing guidance would then be used until impact with the target or detonation of a proximity-fixed warhead.

Hybrid Guidance

A combination of command guidance and semi-active homing guidance is a type of hybrid guidance. It achieves many advantages of both systems. It attains long-range capabilities by maintaining the tracking sensors on the delivery vehicle (ship, aircraft, or land base) and transmitting the data to the missile. By having the missile compute its own attitude adjustments, the entire mechanization of the fire control problem can be simplified.

Self-contained Guidance Systems

The self-contained group falls in the second category of guidance system types. All the guidance and control equipment is entirely within the missile. Some of the systems of this type are: preset, terrestrial, inertial, and celestial navigation. These systems are most commonly applicable to surface-to-surface missiles, and electronic countermeasures are relatively ineffective against them since they neither transmit nor receive signals that can be jammed.

Preset Guidance. The term preset completely describes one guidance method. When preset guidance is used, all of the control equipment is inside the missile. This means that before the missile is launched, all information relative to target location as well as the trajectory the missile must follow must be calculated. After this is done, the missile guidance system must be set to follow the course to the target, to hold the missile at the desired altitude, to measure its air speed and, at the cor-rect time, cause the missile to start the terminal phase of its flight and dive on the target.

A major advantage of preset guidance is that it is relatively simple compared to other types of guidance; it does not require tracking or visibility.

An early example of a preset guidance system was the German V-2, where range and bearing of the target were predetermined and set into the control mechanism. The earliest Polaris missile was also designed to use preset guidance during the first part of its flight, but this was soon modified to permit greater launch flexibility.

The preset method of guidance is useful only against stationary targets of large size, such as land masses or cities. Since the guidance information is completely determined prior to launch, this method would, of course, not be suitable for use against ships, aircraft, enemy missiles, or moving land targets.

Navigational Guidance Systems. When targets are located at great distances from the launching site, some form of navigational guidance must be used. Accuracy at long distances is achieved only after exacting and comprehensive calculations of the flight path have been made. The mathematical equation for a navigation problem of this type may contain factors designed to control the movement of the missile about the three axes-pitch, roll, and yaw. In addition, the equation may contain factors that take into account acceleration due to outside forces (tail winds, for example) and the inertia of the missile itself. Three navigational systems that may be used for long-range missile guidance are inertial, celestial, and terrestrial.

Inertial guidance. The simplest principle for guidance is the law of inertia. In aiming a basketball at a basket, an attempt is made to give the ball a trajectory that will terminate in the basket. However, once the ball is released, the shooter has no further control over it. If he has aimed incorrectly, or if the ball is touched by another person, it will miss the basket. However, it is possible for the ball to be incorrectly aimed and then have another person touch it to change its course so it will hit the basket. In this case, the second player has provided a form of guidance. The inertial guidance system sup-plies the intermediate push to get the missile back on the proper trajectory.

The inertial guidance method is used for the same purpose as the preset method and is actually a refinement of that method. The inertially guided missile also receives programmed information prior to launch. Although there is no electromagnetic con-tact between the launching site and the missile after launch, the missile is able to make corrections to its flight path with amazing precision, controlling the flight path with accelerometers that are mounted on a gyro-stabilized platform. All in-flight accelerations are continuously measured by this arrangement, and the missile attitude control generates corresponding correction signals to maintain the proper trajectory. The use of inertial guidance takes much of the guesswork out of long-range missile delivery. The unpredictable outside forces working on the missile are continuously sensed by the accelerometers. The generated solution enables the missile to continuously correct its flight path. The inertial method has proved far more reliable than any other long-range guidance method developed to date.

Celestial Reference. A celestial navigation guidance system is a system designed for a predetermined path in which the missile course is adjusted continuously by reference to fixed stars. The system is based on the known apparent positions of stars or other celestial bodies with respect to a point on the surface of the earth at a given time. Navigation by fixed stars and the sun is highly desirable for long-range missiles since its accuracy is not dependent on range. Figure sketches the application of the system as it might be used for a guided missile.

The missile must be provided with a horizontal or a vertical reference to the earth, automatic star-tracking telescopes to determine star elevation angles with respect to the reference, a time base, and navigational star tables mechanically or electric-ally recorded. A computer in the missile continuously compares star observations with the time base and the navigational tables to determine the missile's present position. From this, the proper signals are computed to steer the missile correctly toward the target. The missile must carry all this complicated equipment and must fly above the clouds to assure star visibility.

Celestial guidance (also called stellar guidance) was used for the Mariner (unmanned spacecraft) interplanetary mission to the vicinity of Mars and Venus. ICBM and SLBM systems at present use celestial guidance.

Terrestrial Guidance Methods

Prior to micro-miniaturization of computer circuits, the various methods of terrestrial guidance proposed had significant limitations. These proposed early systems included an inertial refer-ence system, a television camera to provide an image of the earth's surface, and a film strip of the intended flight path. The guidance system would compare the television picture with the projected film strip image and determine position by matching the various shadings in the two images. This method proved too slow in providing position data, even for a sub-sonic missile. Its other distinct disadvantage was that it required extensive low-level aerial photography of each potential missile flight path. The danger to flight crews and loss of the element of surprise involved in extensive pre-strike photo reconnaissance made such a system impractical.

With the availability of compact mass memory and vastly in-creased computational capability compatible with missile space and weight limitations, terrestrial guidance methods became practical. The advent of small radar altimeters of high precision provided an alternative to photographic methods with the added advantage that weather and lighting conditions were relatively inconsequential. The radar altimeter provides a coarse means of detecting surface features by their height, which can then be compared with stored data concerning expected land contours along the missile flight path. The missile guidance system contains expected land-elevation values to the left and right of the missile's intended ground track. The guidance system will determine that the missile is located at a position where the stored data most closely matches the observed altitudes as pictured in figure. Once the direction of turn and the distance required to correct the error have been determined, the missile will turn to resume the intended track. This method is called Terrain Contour Matching or TERCOM. Even the most capable TERCOM system has insufficient memory to perform contour matching throughout a flight path of several hundred miles. Therefore, the missile will be provided with a series of small areas known as TERCOM maps along the route to the target. The number of TERCOM maps and their separation is determined by the quality of in-formation available on the area and the accuracy of the missile's inertial navigation system. Sufficient data is available from various sources to support TERCOM such that aerial reconnaissance of most target areas is not required prior to the engagement. TERCOM has sufficient accuracy to find, for example, a large military base within a region; however, it could not provide the accuracy to hit a specific section of that base, such as a group of hangars at an airfield. For this reason, a missile using some variation of TERCOM only would require a nuclear warhead.

Delivery of a conventional high-explosive warhead requires precision that can only be provided by some form of optical device in the terminal stage of flight. A cruise missile flies at altitudes and ranges that would prevent transmission of images back to the launch point. Advances in digitized imagery permit computer storage of grey-shaded scenes in the vicinity of the target. The digitized scene can be compared to data from a television camera in the missile and values of grey shading matched to determine actual position relative to desired position. The missile can correct its flight path to that desired and even finally pick out its target. This method, called Digital Scene Matching Area Correlator or DSMAC, is sufficiently accurate to permit the use of a conventional high-explosive warhead. The DSMAC technique would be used only for the last few miles to the target, with the TERCOM method being used for the majority of the flight path. Both of the above methods are limited by the accuracy of information used to create the digital TERCOM maps and DSMAC scenes that are loaded in the missile's memory. Building and formatting these data files for cruise missiles requires considerable support facilities and talented personnel.

Guided Flight Paths

A guided missile is usually under the combined influence of natural and man-made forces during its entire flight. Its path may assume almost any form. Man-made forces include thrust and directional control as shown in figure. The vector sum of all the forces, natural and man-made, acting on a missile at any instant, may be called the total force vector. It is this vector, considered as a function of time in magnitude and direction, that provides velocity vector control. Paths along which a guided missile may travel may be broadly classified as either preset or variable. The plan of a preset path cannot be changed in mid-flight; the plan of a variable path is altered according to conditions occurring during flight.

Preset Flight Paths

Preset flight paths are of two types: constant and programmed.

Constant: A preset guided missile path has a plan that has been fixed beforehand. This plan may include several different phases, but once the missile is launched, the plan cannot be changed. The phases must follow one another as originally planned. The simplest type of preset guided missile path is the constant preset. Here, the missile flight has only one phase.

The term constant preset may be broadened to include flights that are constant after a brief launching phase that is different in character from the rest of the flight. During the main phase of a constant preset guided-missile flight, the missile receives no control except that which has already been built into it. How-ever, it receives this control throughout the guided phase of flight. Often it is powered all the way. The nature of a constant preset guided-missile flight path depends on how it is powered, and the medium through which it travels.

A torpedo fired from a submarine to intercept a surface tar-get, figure, may describe a straight line--a constant preset guided path.

Programmed: A missile could be guided in a preset path against a fixed target; the joint effect of missile power and gravity would then cause the path to become a curve. A missile following a preset path may be guided in various ways-for instance, by an autopilot or by inertial navigation. The means of propulsion may be motor, jet, or rocket. A more complex type of preset path is the programmed preset. Here, the weapon flight has several phases: for example: a torpedo, as illustrated in figure, executing a search pattern. During the first phase, the torpedo, having been launched in some initial direction other than the desired ultimate direction, gradually finds the desired direction by control mechanisms such as gyros and depth settings. The torpedo then maintains this direction for the remainder of this first phase, at the end of which it is presumed to be in the neighborhood of a target. During the second phase, the torpedo executes a search pattern, possibly a circular or helical path.

Variable Flight Paths

The guided flight paths of greatest interest are those that can vary during flight. In general, the heading of the weapon is a function of target position and velocity. These parameters are measured by continuous tracking, and the resultant missile flight path is determined, assuming that the target motion will remain unchanged until new data is received. There are four basic types of variable flight paths in common use: pursuit, constant-bearing, proportional navigation, and line of sight.

Pursuit. The simplest procedure for a guided missile to follow is to remain pointed at the target at all times. The missile is constantly heading along the line of sight from the missile to the target, and its track describes a pursuit path with the rate of turn of the missile always equal to the rate of turn of the line of sight. Pure pursuit paths are highly curved near the end of flight, and often the missile may lack sufficient maneuverability to maintain a pure pursuit path in the terminal phase of guidance. When this is the case, the missile can be designed to continue turning at its maximum rate until a point is reached where a pursuit course can be resumed. The most common application of the pursuit course is against slow-moving targets, or for missiles launched from a point to the rear of the target.

Pursuit: Lead or deviated pursuit course is defined as a course in which the angle between the velocity vector and line of sight from the missile to the target is fixed. For purposes of illustration, lead angle is assumed to be zero, and only pure pursuit is described. (M =.

Constant Bearing. At the opposite extreme to a pursuit path is a constant-bearing or collision path. The missile is aimed at a point ahead of the target, where both the missile and target will arrive at the same instant. The line of sight to this point does not rotate relative to the missile. The missile path is as linear as the effect of gravity and aerodynamic forces allow. If the target makes an evasive turn or if the target's velocity changes, a new collision course must be computed and the missile flight path altered accordingly. The outstanding feature of this course is that for a maneuvering constant-speed target, the missile lateral acceleration never exceeds the target's lateral acceleration. The major drawback lies in the fact that the control system requires sufficient data-gathering and processing equipment to predict future target position.

Constant Bearing: A course in which the line of sight from the missile to the target maintains a constant direction in space. If both missile and target speeds are constant, a collision course results.

$$\frac{d}{dt} = 0$$

Proportional Navigation: The more advanced homing missiles will employ some form of proportional navigation. The missile guidance receiver measures the rate of change of the line of sight (LOS) (bearing drift, if you will) and passes that information to the guidance computer, which in turn generates steering commands for the autopilot. The missile rate of turn is some fixed or variable multiple of the rate of change of the LOS. This multiple, called the navigation ratio, can be varied during missile flight to optimize performance. A missile employing this method is said to use proportional navigation ratio may be less than 1:1 early in the flight to conserve velocity and increase range. As the flight proceeds, the navigation ratio will in-crease to 2:1, 4:1, or even more to ensure that the missile will be agile enough to counter target maneuvers in the terminal phase of flight.

Proportional: A course in which the rate of change of the missile heading is directly proportional to the rate of rotation of the line of sight from missile to target.

$$\frac{dM}{dt} = \frac{K\,d}{dt} \text{ or M} = K$$

Line of Sight: Here, the missile is guided so that it travels along the line of sight from the launching station to the target. This is, of course, the flight path flown by a beam-riding missile. An alternative form of a beam-riding path is the constant lead angle path. Here the beam that the missile follows is kept ahead of the line of sight by a constant offset. The major advantages of the line of sight path are its flexibility and the minimal complexity of the equipment that must be carried in the missile, since the major burden of guidance is assumed at the launching station.

Line of Sight: Defined as a course in which the missile is guided so as to remain on the line joining the target and point of control. This method is usually called "beam riding."

PERMISSIONS

All chapters in this book are published with permission under the Creative Commons Attribution Share Alike License or equivalent. Every chapter published in this book has been scrutinized by our experts. Their significance has been extensively debated. The topics covered herein carry significant information for a comprehensive understanding. They may even be implemented as practical applications or may be referred to as a beginning point for further studies.

We would like to thank the editorial team for lending their expertise to make the book truly unique. They have played a crucial role in the development of this book. Without their invaluable contributions this book wouldn't have been possible. They have made vital efforts to compile up to date information on the varied aspects of this subject to make this book a valuable addition to the collection of many professionals and students.

This book was conceptualized with the vision of imparting up-to-date and integrated information in this field. To ensure the same, a matchless editorial board was set up. Every individual on the board went through rigorous rounds of assessment to prove their worth. After which they invested a large part of their time researching and compiling the most relevant data for our readers.

The editorial board has been involved in producing this book since its inception. They have spent rigorous hours researching and exploring the diverse topics which have resulted in the successful publishing of this book. They have passed on their knowledge of decades through this book. To expedite this challenging task, the publisher supported the team at every step. A small team of assistant editors was also appointed to further simplify the editing procedure and attain best results for the readers.

Apart from the editorial board, the designing team has also invested a significant amount of their time in understanding the subject and creating the most relevant covers. They scrutinized every image to scout for the most suitable representation of the subject and create an appropriate cover for the book.

The publishing team has been an ardent support to the editorial, designing and production team. Their endless efforts to recruit the best for this project, has resulted in the accomplishment of this book. They are a veteran in the field of academics and their pool of knowledge is as vast as their experience in printing. Their expertise and guidance has proved useful at every step. Their uncompromising quality standards have made this book an exceptional effort. Their encouragement from time to time has been an inspiration for everyone.

The publisher and the editorial board hope that this book will prove to be a valuable piece of knowledge for students, practitioners and scholars across the globe.

INDEX

www.ingramcontent.com/pod-product-compliance
Lightning Source LLC
Chambersburg PA
CBHW080404190526
45161CB00003B/126